Library of
Davidson College

MOST PROBABLE POSITION

MOST PROBABLE POSITION

A HISTORY OF AERIAL NAVIGATION TO 1941

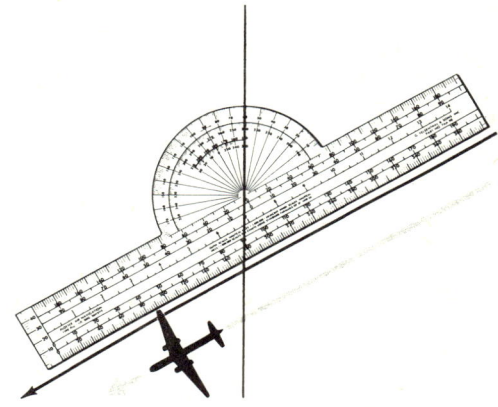

by
Monte Duane Wright

THE UNIVERSITY PRESS OF KANSAS
Lawrence/Manhattan/Wichita

© Copyright 1972 by The University Press of Kansas
Standard Book Number 7006-0092-2
Library of Congress Catalog Card Number 72-79318
Printed in the United States of America
Designed by Fritz Reiber

Preface

I have tried to write, not a pocket encyclopedia, but a general explanation of the problems of aerial navigation and the solutions found to those problems in the pre-electronic era. May the buff forgive my omitting or treating too cursorily his favorite flight or instrument. The reader with a minimum knowledge of navigation may ease his task by skipping Chapter 1, "Marine Navigation before World War I." In the body of the work, I have tried to deal with both civil and military aspects of aerial navigation on an international scale, except for Chapter 7, which concentrates on the U.S. Army Air Corps between the World Wars.

The principal sources on which this book rests are: American, British, French, and German aviation periodicals; documents and reports of those governments and of international aviation bodies; books by participants; the correspondence of Captain P. V. H. Weems (USN, Retired) and of his company, the Weems System of Navigation; textbooks, correspondence, and instructor notes of the Air Corps Tactical School; and the files of the Plans Division, Office of the Chief of the Air Corps.

My obligations are many: to my employer, the United States Air Force (and therefore the American taxpayer), for financing both my graduate degrees; to the graduate history faculty of Duke University, who have achieved a genuine sense of cooperation with each other and with their students; to my dissertation supervisor, Professor Theodore Ropp, whose example has convinced me that a philosophy of permissiveness has a place in higher education; to the librarians at Duke (particularly those of the Main Reading Room and the School of Engineering) and at the Air Force Academy, whose helpfulness I appreciated all the more after working a few days at our country's largest library; to Captain Weems, who opened his home and his files to me. The manuscript has been read in part by Captain Weems and Major General Norris B. Harbold (USAF, Retired); in whole by three of my colleagues, Major David MacIsaac, Major Paul T. Ringenbach, and Captain Edward P. Brynn; and many, many times by my wife. They have saved me from many errors factual and stylistic.

M. D. W.
USAF Academy, Colorado
1 December 1971

Contents

Preface	v
List of Illustrations	ix
1. Marine Navigation before World War I	3
2. The Navigation of Balloons	17
3. Navigation of Airships and Airplanes before World War I	39
4. Navigation in World War I	63
5. Overland Airlines and Radio Navigation	103
6. Transoceanic Airlines and Celestial Navigation	131
7. Navigation of Long-Range Bombers in the U.S. Army Air Corps to 1941	169
Appendixes	
A. Glossary of Abbreviations and Technical Terms	203
B. Flights Referred to in Chapter 6	208
Notes	211
Bibliography	247
Index	271

Illustrations

FIGURES

1. The running fix 5
2. The 45–90 running fix 5
3. Compass rose in use before World War I 6
4. True heading, magnetic heading, and compass heading 8
5. The celestial triangle 12
6. Latitude by meridional transit of the sun 14
7. Computing groundspeed 25
8. The wind triangle 40
9. Two ways to steer an aircraft to a visible point 41
10. Computing the wind from drifts on successive headings 79
11. Direction-finding stations used by Zeppelins 90
12. Removing the 180° ambiguity from a radio bearing 123
13. Radio-range station 124
14. The radio beam 124

PHOTOGRAPHS

Checking compass deviation on an AT-7 navigation trainer 44
Anthony G. Fokker on one of his airplanes 54

Curtiss flying boat *America*	59
Interior of Fokker Tri-Motor	61
Boeing Clipper	61
A collection of navigation computers	68
RAF course and distance calculators	72
Six early U.S. Navy aircraft compasses	74
Aperiodic compass installed in Fokker Tri-Motor	75
Airplane equipped with wireless	81
Two-way radio-telegraph on Glenn L. Martin airplane	82
Zeppelin Staaken giant bomber	94
Handley Page V/1500	95
Model of instrument board used in DH-4	101
Handley Page 0/400 night bomber converted to airliner	104
Bureau of Standards bubble sextant	108
Bygrave slide rule	112
Drift lines painted on Fokker Tri-Motor	112
CAA beacon	119
American Airlines navigational computer	128
The earth inductor compass	135
Instrument board of *Spirit of St. Louis*	136
The Gatty drift indicator	140
Instrument board of Douglas BT-2B	144
Lt. Commander Weems using a marine sextant	146

Colonel Thurlow taking a celestial observation	151
Link bubble sextant	152
Using a sextant in a Fokker Tri-Motor	154
The AT-7 navigation trainer	157
The *American Nautical Almanac* and associated tables	158
Lindbergh's navigation equipment	159
Tables and computations for reducing a celestial sight	161
Route of the Post–Gatty world flight	161
Navigator's cockpit of the Breguet *Question Mark*	163
The Hagner position finder	164
Navigation gear stored in a U.S. Navy O2U	170
Dalton Mark VII navigation computer	182
Ground school for celestial navigation	186
The "Navitrainer" at Barksdale Field	190
Multi-purpose navigation trainer	190
A cadet working a map-reading problem in a Link trainer	191
A cadet shooting a star in a Link trainer	191
Students boarding AT-7 at Kelly Field	192
Interior of the AT-7	192
Navigator at his station in a B-17	195

MOST PROBABLE POSITION

Wenn wir nur eine Ahnung hätten, wo wir sind.

—Log of the balloon *Plauen*,
14 October 1908

MARINE NAVIGATION BEFORE WORLD WAR I

1

Navigation is a hybrid, applied science that uses a variety of instruments and techniques to determine three things: the present position of a vehicle, the direction to steer to reach a desired position, and the estimated time of arrival at that desired position. Sea navigation has a long history, which has been well told.[1] By the 20th century, sea navigation was a mature science, though not a stagnant one. The problems of position, direction to steer, and time of arrival were regularly solved with accuracy sufficient to the needs of the day. When men began to travel through the air, they soon faced similar problems.

The free balloon of the 19th century could not be steered, except indirectly by rising or descending into a favorable wind. But as soon as balloonists attempted long flights or flights above clouds and at night, they needed to know present position, particularly in reference to the coasts and mountains. The *dirigeable* (directable) balloon and the heavier-than-air airplane, as soon as they were physically capable of flying out of sight of the landing field, brought with them the additional problems of direction to steer and time of arrival. It was only natural that the navigator of the ship-of-the-air turn to his ocean-going counterpart to borrow instruments and techniques for solving these problems. The story of air navigation is largely one of the transfer of knowledge and skills from one medium to another. It will be necessary, therefore, to refer to marine navigation frequently; and it is convenient to begin the story of air navigation with a description of marine navigation is it was practiced early in the air age, on the eve of World War I.

Marine navigation divides naturally into two parts.[2] When within sight of land, navigation is both easy and crucial: easy because natural or artificial landmarks abound; crucial because when the shore is in sight, the invisible dangers of shoal water are also close by. Navigation of ships in "narrow waters," with landmarks constantly available, is called pilotage. On the high seas, navigation is more difficult, but accuracy is not so important. Lacking landmarks, the navigator keeps a running account of the speed, distance, and direction his vessel travels. By plotting on a chart the distance and direction made good from the last known position, he can find the ship's position whenever necessary. This procedure is called dead reckoning, and the position so determined is a DR position. With precise instruments and careful attention to details, dead reckoning can produce surprisingly accurate results. But minor errors will creep in, and these accumulate with time, so that the ship's position by dead reckoning becomes increasingly erroneous. Periodically, therefore, the navigator computes the ship's position from an external, independent source; and at the beginning of the 20th century, this could only be celestial navigation. The navigation for an ocean crossing was typically: pilotage in the coastal waters; dead reckoning, periodically restarted from positions determined by celestial navigation, until landfall; and then pilotage to the mooring.

Pilotage involves first, seeing (or hearing) and identifying a coastal landmark, and second, determining the ship's precise position with respect to that landmark. The maritime chart is of prime importance for pilotage. By World War I the most important coasts had been carefully surveyed and mariners could purchase the charts. Where natural landmarks were insufficient, buoys, lighthouses, lightships, foghorns, and submarine bells had been installed, and their locations charted. Governments cooperated in maintaining such aids to navigation and regularly exchanged information concerning them. The major maritime nations published serial "Notices to Mariners," giving the most recent changes affecting aids to navigation. To identify a particular natural landmark, the navigator depended first on his dead reckoning. If an island was seen, what island(s) appeared near the DR position on the chart? The relationship of one landmark to another helped to identify both. For example, in how many places near the DR position was there a small island south of a larger one? Sailing directions, prepared by the same governmental agencies that produced charts, included sketches of the more important landmarks, although the reputation of the old-time coastal pilot depended on his ability to differentiate one headland from another by memory. The chart, the "Notices to Mariners," and the sailing directions included the codes by which the artificial landmarks could be identified.

After identifying a particular landmark, several methods were available for determining the vessel's precise position. The bearing of (direction to) an object could be measured by sighting across the top of a compass. By plotting the reciprocal bearing (the opposite direction) from the known object, the navigator had a line of position on his chart; the ship was somewhere on the line. To find exactly where on the line, the distance to the landmark could be estimated visually, but measurement was preferable. If the height of a landmark, such as a lighthouse, was known, the angle it subtended, measured with a sextant, could be used to find the distance by plane trigonometry. Simultaneous bearings on two objects gave two position lines, and the ship would be at their intersection. When running past a single landmark, a "running fix" was often taken. The procedure is illustrated in Figure 1. The bearing of the landmark was plotted (line A).

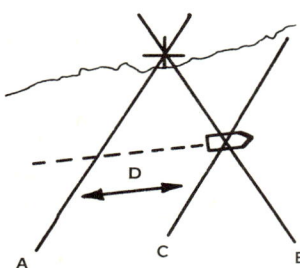 

Fig. 1. The running fix.   Fig. 2. The 45–90 running fix.

At some later time, a second bearing to the same object was plotted (line B), and the first bearing was adjusted to the time of the second, by constructing a line (C) parallel to the original line, and at a distance and direction (D) corresponding to the speed and course of the ship in the interval. A special application of this method was the 45–90 bearing. With the first bearing taken when the landmark was 45° off the nose of the vessel and the second when the landmark was 90° off (which would be abeam, directly right or left), the ship was on the second line at a distance equal to the distance covered by the ship between the bearings. Figure 2 shows the right equilateral triangle used in this procedure. Each leg of the triangle is equal to the time between the bearings multiplied by the ship's speed. Finally, the pilotage charts showed water depth and the composition of the bottom. By taking a depth sounding and sometimes a sample of the ocean bottom, an approximate line of position could be deduced from the chart.

Dead reckoning requires the computation of the direction and distance traveled by the ship. Before World War I three systems of expressing direction were in use in marine navigation. The compass rose in Figure 3 is graduated for all three systems. The oldest of the three was the 32 points of the compass: North, North by East, North-Northeast, etc. A sailing ship that held to one point was well steered indeed, but with the more precise steering possible with steam vessels and better compasses, 32 directions were not enough. Half-points and quarter-points were added. Working with such expressions as "Northwest by West, 3/4 West" was cumbersome and invited error. The second system was then devised, in which the compass rose was graduated in 360 divisions and numbered from North and South, increasing to 90° at both East and West. "NW by W, 3/4 W" then became "North 65° West." A third system was just coming into use. It retained the 360 divisions but counted from 000° at North, clockwise through 360°, which was again North. "N 65° W" became "295°."

Fig. 3. Compass rose showing the three systems for expressing direction in use before World War I. From U.S., Navy Dept., *American Practical Navigator,* p. 15.

Some nations had adopted this system in the late 19th century, but neither the United States nor Great Britain had done so. The U.S. Navy admitted in 1903 that "its general adoption would carry with it certain undoubted advantages" and in 1916 dropped the compass points from its charts.[3] Whichever system of designating direction was used, the instrument for determining direction was invariably the magnetic compass.

The marine compass of 1900 was a brass bowl about 10 inches in diameter, filled with liquid, in which turned a thin, round, metallic card. The card was graduated according to one or more of the directional systems described above. A number of bar magnets were attached to the lower surface of the card and aligned with the N-S direction on the card face. The liquid performed two functions. It lifted part of the card's weight off the supporting pivot, making the compass more responsive to changes in direction; and it damped out or absorbed oscillations that otherwise would have set the card to swinging back and forth. The ship's heading was read opposite the lubber line, a line etched in the side of the bowl, in line with the fore-and-aft axis of the ship. Attached to the top of the compass was an azimuth circle. This device, used for measuring bearings, consisted of a rotatable ring, concentric with the compass, carrying two sighting vanes and a mirror and prism, which brought the compass direction into the field of view when using the vanes. The compass rested on the binnacle, a stand that contained a light and provision for compensating the compass, which will be described below.

A magnetic compass gives true direction only after two corrections have been applied: variation and deviation. A magnetized needle free to turn in the earth's field will align itself with the magnetic meridian, a line of force connecting the earth's two magnetic poles. The magnetic meridians are not straight, however, for the earth's surface is not magnetically uniform. Further, the north magnetic pole is about a thousand miles from the true pole. The angle between the direction of true north and the direction of the magnetic meridian is called variation. Its value can only be determined by accurate surveying. By World War I the marine charts carried lines showing the value of variation over most of the oceans.

A magnetized needle will also respond to local magnetic fields produced by masses of iron and by electrical currents. The error thus introduced into compass readings is called deviation. It can be reduced by careful placement of the compass and by compensation, a procedure in which small magnets and pieces of iron are placed beneath the instrument (in the binnacle), to overcome the local magnetic fields. After compensation, the residual deviation was computed by "swinging the ship." This operation, accomplished while in harbor, involved measuring the compass bearing of a landmark whose magnetic bearing was known and comparing the compass bearing with the magnetic bearing. The difference was deviation. The ship was then moved—swung—to the next compass point and the procedure repeated. The various deviations were recorded on a graph or table, and thereafter all readings of the compass had to be corrected for both deviation and variation.

The continual conversion of true direction to magnetic to compass and back again was a frequent source of mistakes, and practical sailors devised "memory crutches" to aid them. The British sailor learned this rhyme, for use with his deviation graph:

> From compass course, magnetic course to gain,
> Depart by dotted and return by plain;
> But if you wish to steer the course allotted,
> Take plain from chart, and keep her head on dotted.

Figure 4 shows the relationship existing between true, magnetic, and compass directions and should explain why ordinary men needed an

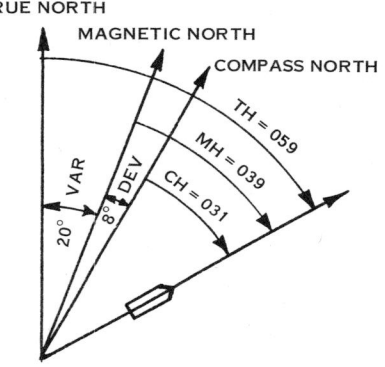

Fig. 4. True heading, magnetic heading, and compass heading.

easily remembered jingle to help them. The problem was universal and the solutions remarkably similar. French sailors memorized:

> Le chien C [compas] court après la viande V [vrai].

Early French aviators were to use it for the same purpose. Likewise, German aviators in World War I were taught:

> Vom wahren zum falschen das falsche Vorzeichen nehmen,
> Vom falschen zum wahren das wahre Vorzeichen nehmen.

At about the same time, U. S. Army aviators learned:

> Variation WEST COMPASS BEST;
> Variation EAST COMPASS LEAST,[4]

which by World War II would have been shortened to

> East is least, west is best.

Along with direction, dead reckoning requires distance, which can be either measured directly or found by first measuring speed. To measure the ship's speed through the water, three methods were available in 1900. The oldest of the three was the chip log, a piece of wood so carved and weighted that it floated low in sea water and acted as a small sea anchor. A light line, attached to the chip log, was knotted at precise intervals, the interval depending on the time-measurement used in conjunction with the log. The United States Navy used a 28-second sandglass, and the knots were spaced 47 feet 3 inches apart, based on the following ratio:

$$\frac{3600 \text{ seconds (1 hour)}}{6080 \text{ feet (1 nautical mile)}} = \frac{28 \text{ seconds}}{47\ 1/4 \text{ feet}}$$

To measure speed, the log was heaved overboard and the glass turned. When the sand ran out, the sailor tending the log counted the number of knots that had run over the side and called out, for example, "Four and one-half knots, sir." Thus "knot" came to mean 1 nautical mile per hour.

A second method of measuring speed was a by-product of steam propulsion. There was an obvious relationship between the revolutions-per-minute of the engines and the speed attained. That relation was determined during a ship's initial trials and could be used as an accurate measurement of speed thereafter, but only for a clean-bottomed ship moving through relatively calm water. As marine growth accumulated on the bottom, or with heavy seas running, the ship's speed fell off, and the navigator either had to allow for this difference by estimate or abandon the use of the RPM indicator to measure speed.

A third method of measuring speed was the pressure log. This instrument, invented in the 1870s, was an open tube, bent in a right angle, with one end under water and facing in the direction of the ship's motion. The height a column of water climbed inside the vertical end of the tube was proportional to the forward speed.[5]

In 1900 the standard instrument for measuring distance directly, instead of first measuring speed, was a development from the chip log called the patent log. The patent or taffrail log was a small propeller pulled behind the vessel on a line suspended from a short boom, so that it rode clear of the wake. The revolutions of the propeller were transmitted to a counter on the ship by means of a flexible cable, and the number of turns was a direct indication of the distance traveled.

Every hour while at sea, the navigator wrote the heading, speed, and distance traveled by the ship in the vessel's log book. Other forces that exerted an influence on the ship's position, but that could not be detected by the instruments described above, were estimated and entered in the log. One such force was the wind. The sideways drift, or leeway, had to be estimated visually. The water was also in motion, carrying the ship with it. The tidal streams in pilotage waters had been carefully charted, and their speed and direction were available to navigators in 1900. At sea, also, the major currents had been charted. Entrance into a current could be detected by the change in water color or temperature; for this purpose, a sample of sea water was taken from a few feet beneath the surface every hour, and its temperature entered in the log. All of these forces—the ship's own motion, the wind, and the movement of the water—acted simultaneously on the ship. In dead reckoning, the effects of these forces were combined mathematically to find the net movement for a given period of time.

Consider the following example of a typical dead-reckoning problem. A ship departed latitude 30° north, longitude 74° west. The patent log read 170 nautical miles (nm). The compass heading was 140°, deviation 2° W, variation 9° E, equivalent to a true heading of 147°. The chart showed the Gulf Stream in that area to be flowing north-easterly with a velocity of 3 knots. The wind was from the north, giving the vessel an estimated leeway of 1 knot to port (left) of its heading, or in the direction of 237°. After an interval of three hours, the patent log read 207 nm. To find the DR position for that time, the navigator first computed the displacement due to each of these forces and converted each to an equivalent displacement north or south, and east or west. The conversion was easily made with the traverse table, which gave the solution of a vast number of right triangles and was found in all the standard works on navigation. The work would be tabulated like this:

	Distance	Direction	\multicolumn{4}{c}{Equivalent distances (nm)}
			N	S	E	W
Ship's motion	37 nm	147		31.0	20.2	
Current	9	045	6.4		6.4	
Leeway	3	237		1.6		2.5
		(add)	6.4	32.6	26.6	2.5
		(subtract)		6.4	2.5	
				26.2	24.1	

The resultant of all three motions was therefore to displace the ship 26.2 nm south and 24.1 nm east of its original position. The southerly movement could be applied directly to the latitude of the original position, for the nautical mile is defined as equal to 1′ of latitude. Therefore the new latitude was

$$30° \; 00′ \; N$$
$$(\text{subtract}) \quad 26.2 \; S$$
$$\overline{29° \; 33.8′ \; N}$$

The length of a minute of longitude, however, varies with the cosine of the latitude, being equal to 1 nm at the equator and decreasing toward the poles. A table of cosines shows that at 40° latitude, 1′ of longitude equals 0.766 nm. The change in longitude was then 24.1 × 0.766, or 18.5′. This value, subtracted from the original longitude because the vessel was traveling on an easterly course,

$$\begin{array}{r} 74°\ \ 00'\ \ W \\ (\text{subtract})\ \underline{18.5\ \ E} \\ 73°\ \ 41.5'\ W \end{array}$$

gave the second coordinate, and the new position could be plotted.

With such a large number of observations, estimations, and computations going into every DR position, it should be evident that slight errors would creep in, no matter how carefully the work was done. It was essential that the accumulated error be eliminated periodically by reference to some object external to the ship. Celestial navigation supplied this need in the years before World War I, as it had for centuries.

The spherical triangle that is at the heart of position-finding by astronomical observation is shown in Figure 5. The triangle may be drawn either on the earth or on the celestial sphere, an imaginary sphere concentric with the earth and of infinite radius. Because spherical trigonometry measures arc as well as angle in degrees, the two triangles are not only similar, they are identical. A point or line on one sphere may be projected onto the other. Two points, the pole and the star (or other astronomical body), are known. The third point, Z, is the observer's position on earth, or his zenith on the celestial sphere; and in navigation this is the unknown that is sought. The two known points define one side, that between the pole and the star. The side between the star and the observer may be found with a sextant, an instrument for measuring angles. The angle measured in this case is the height (or altitude) of the star above the horizon. Since the angular distance from the observer's zenith to his horizon is 90°, the altitude of the star above the horizon, subtracted from 90°, will give the length of the side between the star and the zenith. Thus two sides are known but, as every schoolboy knows, the third side or an angle must also be known before the triangle can be solved.

Referring to Figure 5, if the observer's latitude were known, the side between the observer and the pole could be found. Likewise, if the observer's longitude were known, the angle at the pole could be found. The 19th-century navigator used this possibility. By observing the altitude of a single body and using his best-known (DR) latitude, he would solve the triangle and determine his longitude; or, conversely, using his DR longitude, he would find his latitude. If the navigator used a body nearly due north or south to find latitude, a large error in the DR longitude would cause only a slight error in the computed latitude. Similarly, if a body nearly due east or west was used to find longitude, error in the DR latitude would have little effect. For this reason, early morning and late afternoon observations of the sun were used to find longitude, and a noon observation for latitude.

A different and in many ways preferable way of using the celestial triangle had been discovered in 1837 by Captain Thomas Sumner,

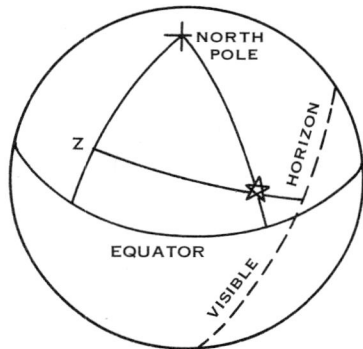

Fig. 5. The celestial triangle.

an American shipmaster. Referring again to Figure 5, consider a number of observers located equidistant from the subpoint of the star. They would all measure the same altitude for that star. Sumner realized that the altitude of a single body defined a circle on the earth, a circular line of position, somewhere on which the observer was located. The usual practice of assuming a value for latitude and solving for longitude (or vice versa) appeared to give a definite position, but the navigator was in a way misleading himself. All that a single observation told him was that *if* he were at the given latitude, *then* his longitude was the resultant value.

The Sumner line, or line of position, was plotted in one of three ways. Sumner discovered the line by working the traditional method three times, using different values of latitude and getting different values of longitude in each case. The three positions so defined lay in a straight line, because the circle of equal altitude is so large that a long arc, often up to 100 miles, approximates a straight line. After the discovery, it was possible to plot the line by working the problem only twice.

A quicker method of finding the Sumner line was to solve the triangle for the angle at the observer. Referring to Figure 5 once more, notice that the angle is the true bearing (azimuth) of the star as seen from the observer's position. The azimuth line may be considered a radius of the circle of equal altitude, and a perpendicular to the radius is tangent to the circle. Again, for considerable distance, the straight-line tangent would not depart appreciably from the circle. To draw the Sumner line by this method, the triangle was solved as before, to find one possible position; then the azimuth was found; and a line perpendicular to the azimuth was drawn through the possible position.

The third method for laying down the Sumner line is called the intercept method. This was introduced by the French Admiral Marc St. Hilaire in 1874. He assumed *both* latitude and longitude for the ship's position and solved the celestial triangle for both altitude and azimuth as they would appear at that position. Then he observed the star and compared the computed with the observed altitudes. The difference he called the intercept. If the observed altitude was greater than the computed altitude, then the observer was closer to the body's subpoint than was the assumed position. St. Hilaire plotted the assumed position, measured the intercept distance in the direction of the subpoint (given by the azimuth), and constructed a line perpendicular to the azimuth; this line was the line of position. If the observed altitude was less than the computed altitude, he measured the intercept distance away from the direction of the subpoint and laid down the perpendicular. One attribute of the intercept method would become of great value later in air navigation: most of the work can be done before observing the body, even before leaving the ground.

However plotted, the Sumner line had one great advantage over older celestial methods. It showed the navigator all the information that could be gleaned from a single observation, and it did not tempt him to see more than was actually there. In spite of this advantage, most marine navigators stuck to the old ways. Rather than having only a line on a chart, they seem to have preferred a result given in the familiar coordinates, even if they could not be certain that it was the correct position. For decades, writers continued to urge the wider use of the Sumner line.[6] The U.S. Navy did so in its official navigation manual of 1903, but at the same time more space was devoted to explanation of the older methods than to the line of position. In 1911 the *Encyclopaedia Britannica* reported that the line of position was "rapidly superseding other modes of ascertaining a ship's position." But in World War II, many merchant marine navigators in the North Atlantic convoys still preferred the old ways.[7]

Two special techniques in celestial navigation were used regularly before World War I. Latitude by Polaris is probably the oldest method of celestial position-finding. Polaris is so close to being directly over the north pole that a single correction, available from a simple table, applied to the observed altitude yields latitude. Equally simple is latitude by meridional transit of the sun. Local noon occurs when the sun is directly north or south of the observer, and noon can be determined by continual observation of the sun, noting the maximum altitude attained. At that instant, the condition illustrated in Figure 6

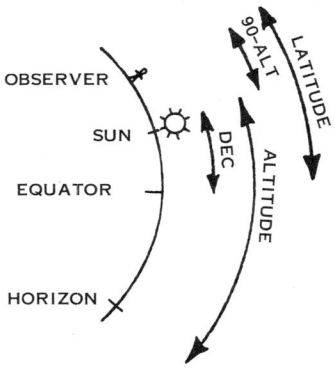

Fig. 6. Latitude by meridional transit of the sun.

exists. The sun's declination (angular distance north or south from the equator) is known. The observed altitude is measured, so that the arc between the observer and the sun's subpoint is also known (90 minus the observed altitude). This last value, applied to the sun's declination, gives the observer's latitude. As with Polaris, a minimum of computation is necessary. The pre-World War I marine navigator regularly took a noon sight for latitude, if the sun was visible.

The equipment needed for celestial navigation was a chronometer, a sextant, a nautical almanac, and a set of mathematical tables. Accurate time is essential for celestial navigation, for the heavenly bodies are in constant motion. (In the matter of relative motions, the navigator finds it convenient to forget Copernicus.) The marine chronometer was the best clock that the ship's owner could afford, and it was treated with great respect. The care of the chronometer almost amounted to a ritual: it was wound with a given number of turns, by the same person, at the same hour each day. It was kept in a special padded case, never taken on deck, and protected against extreme temperatures. It was seldom reset, but allowed to run undisturbed. Every day while ashore, the instrument's reading was compared to accurate time from an observatory and the error recorded. After several weeks the average daily error—the rate—of the chronometer could thus be determined, and the rate was then used at sea to correct the indicated time. The accumulated error might reach several hours during the voyage, but if the rate was approximately constant, accurate time could be found. The advent of the telegraph cable had simplified the problem of timekeeping, because by the 20th century every major port had a telegraph office and access to correct time.

The marine sextant had been virtually perfected by 1900. The optical system incorporated a low-power telescope, which was used to sight on the visible horizon as a reference line. A mirror and a prism reflected the image of the sun (or other body) into the field of view with the horizon. When the body was resting exactly on the horizon, a scale showed the height of the body above the horizon in degrees and minutes. Under ordinary conditions at sea, experienced navigators could measure altitude to the nearest 1/2' of arc, which when reduced gave a position line that was accurate to within 1/2 mile.

To use the sextant, the observer had to be able to see both the body and the horizon beneath the body. Lacking bright moonlight, the night horizon was usually insufficiently distinct, and most observations were taken during daylight hours. During twilight, however, there was a short period when the stars and the horizon were visible, and star observations could thus be taken twice daily, weather permitting.

The sextant did not measure the true altitude of the body; as with most navigational instruments, a series of corrections had to be applied to the sextant altitude to find the true value. First, there was apt to be an error in the linkage between the mirror and the scale in the sextant. This error had to be measured and then a correction applied to all readings. In observing the sun or moon, it was usually more convenient to measure the altitude of the top or the bottom of those bodies, but the theory of celestial navigation dealt with their centers. A correction for the radius of each body had to be applied. The locations of the navigational bodies were tabulated as though they were observed from the center of the earth. The observer on the earth's surface measures a slightly different angle; for the sun and moon, this error, called parallax, was large enough to require correction. Light rays from space are refracted, or bent, on entering the earth's atmosphere. This too had to be corrected. Finally, an observer on a ship's deck sees the horizon farther away than would be the case were his eye at sea level. This error, called dip, also had to be corrected. The effect of all these errors, combined arithmetically and applied to the sextant altitude, gave the true altitude, and the navigator could proceed with the solution of the celestial triangle.

One corner of the triangle, it will be recalled, is the celestial body. By 1900 each major maritime nation was producing an annual nautical almanac, the function of which was to provide the exact position of the sun, moon, visible planets, and navigational stars at any particular time. The coordinates used to locate bodies on the celestial sphere were declination and right ascension. Declination, the angle north or south of the equator, corresponded to latitude. The east-west coordinate, right ascension, was measured eastward from the "first point of Aries," which is the point where the sun crosses the equator northward at the vernal equinox. The almanacs tabulated the locations of the navigational bodies at noon, Greenwich mean time, for every day. The rate of change for both coordinates was also given, so that navigators could compute the exact positions for other times.

Each ship carried a set of mathematical tables, many of which (the traverse table, table of tides, the trigonometric functions of natural numbers, the correction for latitude by Polaris) have already been mentioned. The astronomical triangle was solved, like any other spherical triangle, with the aid of a table of logarithms of the trigonometric functions. Twenty to 30 other miscellaneous tables were included, along with completed sample solutions of all the various problems the navigator had to work.

Both celestial navigation and dead reckoning were subject to observational and computational errors. Both required that a complex procedure be followed meticulously. Fortunately, the physical conditions under which the navigator worked were satisfactory. Every naval vessel, at least, had a well-stocked chart room, containing a large, well-lighted table equipped with the necessary plotting instruments: dividers, draftsman's compass, parallel rules, protractor. All naval officers knew the theory of navigation, and thus a colleague could be asked to give an opinion or to check a computation. There was also enough time to do careful work, although ship speeds were increasing. A Royal Navy navigation instructor cautioned in 1911 that, with steam vessels moving at 20 knots, the navigator must work quickly; "otherwise when laid down on the chart the position shows where the ship *was,* and not where she *is.*"[8] All of these conditions—working space, trained assistants, a relatively leisurely pace—would change for the worse, as soon as the instruments and techniques of marine navigation were introduced into early aircraft.

THE NAVIGATION OF BALLOONS

The story of aerial navigation begins with the free balloon. Hot air balloons and gas balloons, both large enough to lift men, were invented in the 1780s. The succeeding 20 years saw a flurry of interest in ballooning, including some scientific experimentation and the first attempts to use the balloon in war. An overnight voyage of 500 miles and an ascent to 23,000 feet were achieved. For the next 30 years, balloon flights were sporadic and no particular progress was made. Then, in the middle of the 19th century, mechanical power was applied with limited success to propel an oblong balloon. Thereafter interest in the free balloon for its own sake was reinforced by the greater potential inherent in the airship. A number of scientific organizations took up ballooning as a means to conduct experiments, enthusiasts founded aero clubs which published specialist journals,[1] races were organized, money came from wealthy benefactors and governments. Ballooning, in short, became institutionalized. The various problems relating to the balloon and its uses were investigated systematically. Flight became less a stunt, more an accepted sport with possible commercial and military applications. Against this background of increasing activity, with more frequent and longer balloon flights, the ability to determine position came to be recognized as a practical problem. Before the balloon's place was finally usurped by powered aircraft, aerial navigators had devised and used a surprising number of instruments and techniques.

In a balloon, the most obvious way to find one's position is to look down at the ground beneath. The first, unconscious method of aerial navigation, as it had been the first method of marine navigation, was direct observation of the ground. As soon as the balloon drifted away from the familiar area surrounding the departure point, a map was needed. Keeping track of one's position in the air by map reading is not as simple as it might appear, unless the observer can give the problem his undivided attention. Balloonists, however, were often occupied with experiments, taking photographs, or simply sight-seeing. Further, orientation was complicated by the tendency of the balloon to rotate slowly, but continually, about its vertical axis. Thus the relationship of the observer to any point on the surface never remained constant for long. To determine the track of the balloon across the ground, it was advisable to identify a point directly beneath the balloon and mark it on the chart. Sometime later, if a succeeding position was likewise identified and marked, the track made good was obvious. By noting the time of the two positions, the groundspeed of the balloon could be computed easily. Then, by assuming that the wind would remain constant, the balloonist could project the position of the balloon forward for any interval of time. This procedure helped to locate a subsequent position, because the observer's attention was directed to a relatively small area on the chart before he started looking for a distinctive landmark.[2]

To identify a landmark on the earth by comparing it to a chart, the particular landmark must, of course, be on the chart; and it must be represented in such a way that it can be identified from above. It may not be easy to realize today, when aerial photographs are so common, how different the surface of the earth appears when first viewed from above. Numerous maps were available in the 19th century, but none of them had been prepared for aerial navigation. Cities and towns were ordinarily represented by symbols that classified them by population. To identify the city from above, the outline of the built-up area is more useful. Railroads show distinctly from the air, and balloonists sometimes carried railroad maps. These maps showed the cities served by each line and identified each station, but there was no particular attempt made to indicate where and how the railroad curved, or the angle at which a railroad left a city, or the angles made by roads intersecting railroads. (Balloonists sometimes carried railroad timetables, in addition to maps, to know, when passing a station, "when there is a train to take one back, or whether to go on to another station on the line of advance.") Military maps were the best available for the new purpose, for general staffs had long been concerned with precise topographical detail; but not having been prepared for aerial use, they did not generally show the earth's surface as it appeared from above.[3]

The first published plea for an aeronautical chart was apparently that of Hermann W. L. Moedebeck, a Prussian artillery officer and balloonist, in 1888. He pointed out the difficulty of determining the nature of the ground when flying over certain areas, such as cultivated, hilly country. It was therefore difficult to choose a safe landing place, protected from the surface winds. Charts designed for air use could overcome this difficulty by using distinctive symbols to indicate safe landing places.[4] The suggestion was premature, for productive work on aeronautical charts was not to begin for 21 years.

The 2d International Congress of Aeronautics (a private organization composed of representatives of the national aero clubs and dedicated to the furtherance of all aspects of flight), meeting in Paris in 1900, appointed a subcommission to investigate the problem of position finding in a balloon. The subcommission considered the publication of a special map, but its attention was soon diverted to an allied topic: a scheme to classify the principal topographic elements shown on existing charts and catalog them systematically. A balloonist then would need only to note the major topographic features over which he was flying, consult a set of tables prepared to accompany the chart he was using, and extract coordinates that would be within 1 mile of the balloon's actual position. This proposal might be feasible with an electronic computer, but the subcommission decided it lacked the time and money to execute the proposal and dropped it. The members next interested themselves with means to find the position of an aircraft within and above clouds and did not return to the idea of publishing a special aeronautical chart.[5]

In 1906 Moedebeck took up the subject again. Admitting that charts prepared for other purposes had served the balloonist in the past, he pointed out several conditions under which position determining was particularly difficult. With the increasing duration of balloon flights, including flights at night and through clouds, the time had come to produce a chart designed for the task. The chart should include those things most useful to the balloonist and airship navigator. Landmarks should be represented by symbols designed for quick and positive identification. The cartographer should pay particular attention to those things that would be visible at night: all towns, bodies of water, even railroad signals. At night the balloonist not only looked to the earth for clues; he also listened. "The rustling of woods, the rolling of waves, the chugging and whistling of trains, the hubbub of the big city, the pounding and hammering of industrial regions, the animal voices from the villages," even the frogs of the swamps were useful.[6] The sources of characteristic sounds should therefore be charted. As in his 1888 article, Moedebeck was still considering how charts could contribute to safe landings. Electric power lines should certainly be included; twice in December 1905 balloons had drifted into power lines, with disastrous results. The editor of *Aéronautique* agreed with this point and published a map of French power lines in 1907; Moedebeck used the data in preparing later charts.[7]

Moedebeck recognized the difficulties of the task he was proposing and anticipated that it would be uneconomical for private firms to undertake it. Government support seemed essential. One means of governmental assistance, which would also make air charts available quickly, would be to overprint aeronautical symbols on existing military topographical charts. He was soon to have an opportunity to put his ideas into practice. In 1907 the 3d Congress of the International Aeronautical Federation (a rival of the International Congress of Aeronautics; the two organizations pursued similar ends and had several common members), meeting in Brussels, formed an International Commission for Aeronautical Charts, of which Moedebeck was appointed chairman. He was also chairman of the German branch of the International Commission, the *Deutsche Luftschifferverband,* which produced the first aeronautical chart.[8]

He chose, as the basis of his work, a new series of topographical charts being produced in the Cartographic Division of the Royal Prussian Land Survey. This chart, when completed, would cover middle Europe from southern Norway to northwestern Rumania, and from London to Smolensk, on a scale of 1:300,000 (about 5 miles/inch). The sheets covering northern and central Germany, Denmark, and the Low Countries were already available. He chose the Cologne sheet as a prototype because, containing both industrial areas and moderately high terrain, it would be a fair test of his method of overprinting aeronautical symbols (such as landing areas, gas production and storage facilities, coastal lighthouses, lines of magnetic variation, hazards to flight) on existing charts. He indicated terrain elevation by a solid color overprint, one color for each 250 meters up to 1,000 meters, thereafter every 500 up to 3,000 meters. These division points were chosen to correspond to the levels for which weather observations were taken. Further, contour lines were drawn at 50, 100, 200, 300, and 400 meters, for the benefit of airships and airplanes, both of which still flew at very low altitudes.[9]

The Cologne sheet was shown at the Frankfurt a. M. International Aeronautical Exposition in the fall of 1909. The German Commission approved the design, and Moedebeck was authorized to produce additional sheets as finances allowed. He intended to complete the Berlin and Lake Constance sheets next, one being an example of a perfect plain and the other of the highest mountains. Then the Commission could judge whether the product met the need. If so, subsequent efforts should be directed toward those sheets needed by the new Zeppelin transport company.[10]

From this point on, the story of aeronautical charts is more complex. Additional organizations and individuals became interested, and the literature proliferated in proportion to the differences of opinion. To confuse matters further, the chart makers tried to meet the needs of airplane and airship pilots, though there was little experience to go on, while satisfying the better-established requirements of balloonists.

A commercial firm offered an air chart of Paris in the summer of 1909. The direction and distance to various French cities were shown in the border, and aeronautical symbols were included. In September the 4th International Congress of Aeronautics moved that "an international meeting be convoked to agree on the conventional signs for use on aeronautical charts." In 1910 the Aero Club of France organized a cartographic commission. By the end of 1911 it had produced sample sheets and received a subsidy from the Ministry of Public Works.[11] Concurrently the French army had decided, after the maneuvers in Picardy in 1910, that topographical maps were unsuitable for military aviation. The army's Geographic Service then overprinted aeronautical data on existing charts and, in time for the maneuvers of 1911, published the four sheets needed to cover the area of operations. The British General Staff adapted an Ordnance Survey map for air use in the maneuvers of 1911, and the British Association considered the matter of air charts in September of the same year. The next month the first air chart appeared in the United States, when the Automobile Blue Book Publishing Company produced an air map of western Long Island for the Aero Club of America.[12]

The International Commission for Aeronautic Maps, the German branch of which had produced the first map, tried to establish international standards for air charts in May 1911. Representatives from most of the western European nations agreed to standardize their individual charts on the 1:200,000 scale (about 3 miles/inch); to use the conformal, conic projection previously chosen as the basis for the International Map of the World; to express position in conventional degrees and minutes; and to follow a common indexing system to identify the many sheets. Beyond that, agreement was almost impossible.[13] The many questions relating to air maps continued to be debated vigorously.

The arguments ranged over virtually every aspect of cartography: scale, sheet size, the symbols to be used, the method of portraying elevation, coloring. There could be no single best solution to any one of them; compromise had to be sought. The larger the scale, the more intelligible detail that could be provided, but also the higher the cost. Larger sheet sizes made the total series cheaper to produce, but increased the price of the individual sheet. Large sheets were also cumbersome to use.[14]

The choice of symbols was largely a matter of individual preference, but again, the larger the number of symbols, the more difficult the chart would be to interpret. Some advocated using aerial photographs directly as charts; this would certainly include all possible detail. A photograph, however, shows so much detail that important features are hidden, and it also introduces the difficult problem of perspective. The need to eliminate excess detail was generally admitted; the problem was to agree on what constituted excess.[15]

Excepting symbols, the method of portraying elevation was perhaps argued longer than any other question. Moedebeck's solution of using colors of extreme contrast was approved by some participants in the German Circuit, an airplane race from Berlin to the Rhine and back in the summer of 1911. Such charts gave an impression of a terraced earth, however, one elevation giving way abruptly to the next, and the harsh color contrasts, after long periods of scrutiny, must have tired one's eyes. There were advocates for using tints that changed only slightly from one elevation to the next. Attempts to color charts so as to represent the appearance of the earth as closely as possible (green woods, blue water, brown open fields) were abandoned, for this prevented the use of color to show elevation.[16]

Attempting to provide a common chart for balloon, airship, and airplane proved nearly impossible, because the various experts offered contradictory advice.[17] But one common chart having proved almost too expensive, three were out of the question.

Finally, the attempt to overprint aeronautical symbols on existing charts, a compromise measure to reduce costs, produced a chart that was almost illegible. Two German military pilots, writing in 1912, dismissed Moedebeck's 1:300,000 chart as "scarcely suitable for airships, not at all for airplanes," because of the clutter. The French charts produced the same way, but using a base chart drawn to a larger scale (1:200,000) were somewhat better, but the industrial areas contained a maze of symbols. They preferred to use German General Staff topographic charts, adding by hand the aeronautical data they felt essential.[18]

This preference for military topographic charts continued into the first years of the war, and fortunately so, for not much of Germany was covered by air charts in 1914. In Britain, the Ordnance Survey charts had no rival. "France alone, among the warring nations, had any considerable number of sheets completed of a basic air map.... They embraced most of the region between Le Havre and the Swiss border." The United States had scarcely begun. The Coast and Geodetic Survey provided topographic maps extending inland three miles from the coast; most of the interior had not been surveyed to topographic standards.[19]

As aeronautical charts improved in quality, so methods for employing them in balloon navigation became more sophisticated. As previously described, the simplest method was to identify a landmark directly beneath the balloon, note the time, and deduce the track and groundspeed from two such observations. If a balloon passed near an identifiable landmark, the balloonist could visually estimate his position with reference to the landmark. In the history of navigation, however, estimate gives way to measurement in the interest of accuracy. Marine navigators used an instrument that allowed precise positioning with reference to any visible, charted landmark, no matter how far away. This instrument was the bearing compass, a magnetic compass fitted with sighting vanes. When the vanes were aligned with an object, the magnetic bearing to that object could be read on the compass rose.

Using a compass in a balloon basket was difficult. Because the balloon continually rotated, the compass continually rotated, as the needle sought north. A movement by anyone in the basket, whether it was the observer moving into position to use the sighting vanes or someone else idly changing his position, tilted the basket and disturbed the compass. Any reading of a compass had, therefore, to be an instantaneous reading, not as accurate as it might be under stable conditions. The difficulty could be overcome partially by using a liquid-damped compass, such as had been designed for small, fast ships. This compass was made intentionally sluggish, so that it would not respond to every slight disturbance. The position in which the compass was mounted also affected its stability. Some models were designed for mounting in the balloon ring, where the lines coming down from the balloon were attached to the basket. This was just above head height for a man standing in the basket. To be read from beneath, such compasses had inverted roses, similar to those installed in submarine periscopes. While this was the most stable position, bearings could not be taken with the compass there. It had to be moved to a bracket mounted on the edge of the basket. Another model was provided with a gimbal mount, suspended from the balloon ring. A pendulous weight beneath the center of the compass improved the stability.[20]

For accurate results with a magnetic compass, balloonists had to consider deviation, the error caused by a mass of iron near a compass. A balloon did not contain much metal—only fittings and instruments; but these were necessarily close to the compass, because of the small size of the basket. Compensation, in the manner of marine compasses, was considered; but it was simpler to banish iron from the basket, using aluminum and brass instead.[21]

With the bearing compass, position could be found by taking bearings on two landmarks and plotting them on a chart. Two such positions could be used to compute track and groundspeed in the same way as positions found with the map alone. The basket provided scant room for spreading out charts and drafting tools, however, and an accurate track could be found more easily by noting an object directly beneath the basket and later taking a bearing on that object. The balloon's average track for the period of time would be the reciprocal of the bearing. For overwater flights, French naval balloonists provided themselves with artificial sighting points by throwing small rubber balloons into the sea. At night they used small phosphorous charges attached to cork floats.[22] All these methods provided track information only after a period of time. There was one method of measuring track instantaneously, but unfortunately the option was seldom available, because it depended on the use of the guide rope.

Efficient operation of a balloon requires maintaining a nearly constant altitude. When a balloon rises, the gas expands and some must be released to prevent rupture of the bag. To halt a descent, ballast must be dropped. When either gas or ballast becomes seriously depleted, the balloon has to land. The guide rope served as an altitude stabilizer. It was simply a long piece of rope trailed along the ground. As the balloon rose, it picked up more rope and automatically increased its effective weight. As it descended, more of the rope's weight was supported by the ground. With a bearing compass, it was a simple matter to sight along the guide rope and determine the track of the balloon. The guide rope, however, tended to snag in fence rows and to drag tiles off roofs. Its use was restricted to flights over water or uninhabited land, as well as to low altitudes. Thus this simple method of determining track was generally unavailable to the balloonist.[23]

In addition to track, the rate of motion along that track is required in navigation. Both items could be found, indirectly and after a period of time, by measurement between two positions on the chart. Track could be found more easily by taking a back bearing on an object that had earlier passed directly beneath the balloon. A faster method of finding groundspeed was also needed. The compass and the use of bearings had been borrowed from naval science, but that source had little to offer for the direct measurement of groundspeed. Naval speed instruments measured the relative motion between the ship and the medium in which it moved, the water. A balloon, however, has no motion relative to the air. It moves only with the wind. Except for the special case of a balloon flying low over water, where a naval log could be towed on the end of a line, new methods had to be sought.[24] Before the days of electronics, the only method found to be generally useful was based on visual measurement of the apparent motion of the earth beneath the aircraft.

Some of the balloons the French flew out of besieged Paris in the winter of 1870-71 were equipped with primitive groundspeed meters. A telescope of known focal length, containing a flat glass with a circle of known radius etched on it, was mounted in a ball-and-socket joint so as to point straight down. By timing any object across the radius of the circle, and knowing the altitude above the ground, the groundspeed could be found from a table supplied with the equipment.[25] The principle behind this instrument, and its many successors, is illustrated in Figure 7.

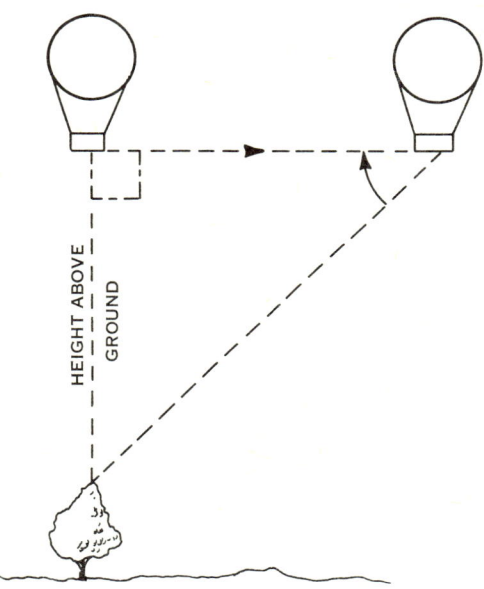

Fig. 7. Computing groundspeed. If one side and one angle of a right triangle are known, the remaining values can be found. In the simplest case, the balloonist timed the apparent motion of an object on the ground from directly beneath the balloon until it was 45° astern. The distance traveled by the balloon in the interval equaled the height of the balloon above the ground.

In 1898 an altogether different method of measuring groundspeed was described. If flying at a low altitude, a small weight attached to a long length of thread was thrown over the side. After allowing enough time for the weight to reach the ground, the rate at which the thread was pulled off its spool could be measured and the groundspeed determined. The inventor admitted that the inertia of the long length of thread retarded the movement of the balloon.[26] Computations would therefore have been too low. If used over inhabited land the procedure also suffered from the same public-relations defect as the guide rope, and it was never widely used.

By 1909 an improved groundspeed meter was in use; like the 1871 instrument, it solved the vertical triangle (Figure 7). A bearing compass, universally mounted, was suspended over a hole in the bottom of the basket. A circle of 1 cm radius was etched around the center of the compass rose, which was transparent. A lens of 10 cm focal length projected a picture of the earth beneath onto the compass rose. By timing the passage of any object from the center of the compass rose until it crossed the circle, the groundspeed could be computed, because the distance covered on the ground was ten times the balloon's height. Where the object crossed the compass rose leaving the field of view also indicated the track of the balloon. A single instrument thus provided both track and groundspeed in less than a minute.[27] For the instruments of 1871 and 1909, and all subsequent variations, the observer's height above the surface of the earth had to be known. It could be found, at least approximately, with the altimeter.

The pressure altimeter has been used in aircraft from the beginning of human flight,[28] and it is still found in the most modern airplanes. The instrument measures atmospheric pressure but is graduated in units of height. Unfortunately, the relationship between pressure and height is not uniform. The pressure at sea level changes from place to place and from time to time. The relationship between pressure and height also varies vertically, because as the temperature of air changes, its pressure changes. A pressure altimeter may be calibrated to give the correct height for any particular rate of change of pressure with height; but whatever rate is chosen, actual conditions are likely to differ, and the indicated height will be erroneous. Even if a pressure altimeter were free from instrumental errors, two things would have to be known to correct the indicated altitude: the pressure existing at sea level and the temperature of the air at flight level. The first can be found before leaving the ground, but the value will be valid only for that location and time. An ordinary thermometer serves to find the second.[29]

The first balloon altimeter was the mercury barometer, taken from the laboratory and set up in the basket. The mercury barometer is about 3-feet tall. It could be used in a balloon, but a smaller instrument was obviously desirable, and several models of aneroid barometer appeared during the 19th century. An aneroid is a small box from which the air has been evacuated. The ends of the box are made of very thin metal, held apart by a spring. The metal flexes as outside pressure changes, and this slight movement is multiplied by a number of pivoted arms, ultimately moving a pointer over a dial indicating height.[30]

Balloonists carried altimeters for several reasons. Scientific experiments in the air required a knowledge of altitude. For long-distance flights, a reasonably constant altitude had to be maintained. For landing at night, knowledge of height was obviously desirable. But when none of these purposes pertained, an altimeter was apt to be carried merely to satisfy a very human curiosity. "There is in every pilot's mind, especially in the beginner's, an unexplainable desire, one might even say a craving, to know how high he is." Recording aneroids that continued to show the maximum height attained until they were manually reset allowed sportsmen to bring back proof of achievement to doubters on the ground.[31]

The aneroid thus became a common instrument, if not a particularly accurate one. When precise altitude measurements were needed, other methods were substituted if possible. For aircraft trials and contests, triangulation from the ground was used to compute exact height. The same triangle could be solved with an aerial photograph, if the picture contained two identifiable points of known distance apart.[32] French naval balloonists suspended two electrical conductors, held closely together by an insulator, below the balloon when low over the sea. If sea water completed the circuit, a bell rang in the basket. One theorist proposed measuring the force of gravity to determine altitude. An acoustical method was proposed: measure the time from the emission of a loud sound to the return of the echo. The pressure altimeter nevertheless remained the only means generally available for measuring height in flight until well after World War I. When the groundspeed meter appeared, requiring the measurement of altitude, it was the pressure altimeter—barometer or aneroid in balloons and airships, the aneroid later in airplanes—that supplied the need.[33]

To compute groundspeed, one had to know the height above the ground. The altimeter showed height above sea level or above the departure point, depending on whether it had been set to read airfield elevation or zero before flight. In either case, the height of the terrain beneath the aircraft had to be considered. This value could be found approximately from the chart. For accurate measurement, the optics of the groundspeed meter had to be held parallel to the ground. To achieve this attitude, the instrument was mounted in such a way that gravity would keep it pointing downward, like a plumb bob. Unfortunately, a plumb bob cannot differentiate between gravity and other accelerations. Because the early groundspeed meters could not be held strictly horizontal, the measurements were in error. This fault did not receive any particular attention in the contemporary literature, and it may not have been widely recognized.[34]

The next instrument of aerial navigation that we will consider, the sextant, also requires an accurate definition of the horizonal. The inability to achieve this definition with precision long threatened to prohibit the use of celestial navigation in aircraft.

To navigate by the methods so far described—map reading, bearings, direct measurement of groundspeed and track—the surface of the earth had to be visible. Yet balloonists on occasion ventured into and above the clouds. When they did so, they had no means for knowing their location or direction of travel. Meteorology of the upper air was an infant science, but balloonists knew that wind currents changed with altitude. The problem was illustrated perfectly in 1899, when a balloon left the Crystal Palace in southern London. The surface wind was easterly. The balloonist ascended through a low overcast, flew at 3,000 feet for two hours, and expected to be near Reading. He then "judged it prudent to dip down through the cloud floor for the purpose of making sure of [his] whereabouts." The balloon was off the east coast; but once beneath the clouds, it again drifted west and reached the land easily. In this case, the winds above and below the clouds were of almost opposite direction.[35]

Keeping track of a balloon's progress at night by map reading was also difficult when the winds changed. In the first Gordon-Bennett balloon race, from Paris in 1906, the balloons first drifted east but by nightfall had reversed direction and were moving slightly south of west. The moon rose, almost full. Then, to follow the account of the navigator on one of the British entries: "When railways below us gave the next opportunity of locating our position, we found none on the map that tallied with those on the ground." After an hour of uncertainty, they crossed a large town. On "the next map to the north we identified the town by its size and the direction of the railways and roads as undoubtedly Evreux." Shouted questions to people on the ground elicited a confirmation. The balloon's direction was now northwest, which made a good run to western England or even Ireland possible. Seven balloons crossed the Channel. The crews of seven others were unable to follow the rapid wind shifts. When they saw a seacoast approaching, they believed it to be the Atlantic Ocean and landed.[36] Thus night map reading over northern France, with a bright moon and no clouds, was not always satisfactory. Yet to win the races, which were judged by distance achieved from departure, the balloonist had to take advantage of the stronger upper-level winds, even if he had to fly above clouds. Celestial navigation, long perfected for use at sea, seemed to be the answer to his navigational needs.

At first, it seemed that celestial navigation might be even more useful for balloonists than for surface navigators. The latter often found the heavenly bodies covered by clouds. This difficulty would not exist for the ballonist, according to an 1892 writer, "because he can always climb above the seas of clouds, the altitude of which rarely exceeds 7,000 feet; above that height . . . only scattered cirrus" would be encountered—an indication of the state of upper air meteorology.[37] Such optimism soon faded when ways and means had to be considered.

To find a position by celestial navigation, the marine navigator used a sextant to measure the height of a star or other body above the horizon. In a balloon, the natural horizon was usually unavailable. Over the sea, unless the balloon was very low, the horizon was indistinct. Over land, it was irregular and not acceptable as a datum plane. At night, it was invisible. The star's height therefore had to be measured above an artificial horizon. Marine navigators had investigated this possibility, because they too needed to be able to take celestial observations when the natural horizon was invisible or indistinct. To this end, various instruments had been proposed, tested, and generally discarded.

One artificial horizon often used with the sextant was a container of mercury. If perfectly still, the surface of the mercury made a more precise horizontal reference than the natural horizon. Naval officers used it for survey work on shore, but it could never be used on shipboard. Other devices incorporating plumb bobs, pendulums, and carpenters' levels were tried and discarded. Only one artificial horizon promised to be useful at sea: the gyroscopic sextant invented in 1885 by the French Admiral Fleuriais.

To an ordinary marine sextant, Fleuriais attached a gyro driven at high speed by a battery-powered air pump. A line drawn on the sides of the gyro provided the horizon. A lens system made the line appear to be focused at infinity and projected the image into the sextant optics. The observer aligned the body with the horizon in the same way as when using the marine sextant. In theory, the instrument was sound; but perfecting it for practical use took many years, and it remained expensive, delicate, and cumbersome. Although the battery was packed in a satchel to be carried by a shoulder strap, the weight of the sextant and gyro required a semirigid support from the operator's belt. The instrument was used successfully in a balloon at least once, but the success was attributed to the skill of the observer and not to the efficiency of the instrument.[38]

The other possible artificial horizons, all of which had been found wanting in marine use, were tried in balloons. Pendulum devices were tried in Germany in 1887 and in France in 1892. Liquid horizons were tried in both countries during the same period. Mercury horizons were found never to "come to rest in balloons. When one is extraordinarily still, the pulse beat alone is sufficient to make the glycerin horizon too unsteady . . . ."[39] One writer suggested "an oily liquid, such as tar."[40] The more viscous the liquid, the less it was disturbed by extraneous accelerations; but if too viscous, it would not respond to gravity either. The most promising of a bad lot of possibilities was the carpenter's level, an air bubble floating on a liquid under a curved glass.

In preparation for his attempted balloon flight from Spitsbergen to the North Pole in 1897, Salomon August Andrée experimented with a bubble sextant, but he was unable to achieve satisfactory results and apparently carried a marine sextant on the flight. The Butenschön firm of Hamburg produced the first successful bubble sextant for air navigation in 1901. By 1903 it had been used in both Germany and France. The instrument consisted of a low-power telescope attached rigidly to a graduated quarter-circle. An arm, pivoted on the telescope and carrying the bubble chamber, moved along the circle. The image of the bubble was reflected into the field of view of the telescope, so that the observer could view the bubble and the celestial body simultaneously. Cross hairs aided the alignment of the two objects. To measure a height, the telescope was pointed directly toward the body. The arm was adjusted until the bubble was centered in its chamber and aligned with the body under the cross hairs. The height of the body could then be read to the nearest degree where the arm crossed the circle. A vernier refined the reading to about 2′ of arc. A later model incorporated a lighting system for night use.[41]

As compared to the marine sextant, the Butenschön sextant had two major faults. First, when the sextant was accidentally jostled during an observation, the star and the bubble moved in opposite directions in the field of view. This fault was corrected in the second generation of bubble sextants, that produced by Ponthus et Therrode in 1907. The bubble chamber was placed on top of the telescope, thus effectively reversing its position in respect to the optics. The body and the bubble then moved in the same direction and at approximately the same speed, when the sextant was disturbed. The third generation of air sextants corrected the second major fault. The marine sextant had always been held horizontal by being pointed at the sea horizon. The body was brought into the optics via an adjustable mirror. Always holding the instrument in the same attitude was a distinct advantage to the observer. The first air sextants had to be pointed at the body and therefore had to be held at various angles. The Schwarzschild balloon sextant corrected this fault by attaching the bubble chamber firmly to the telescope and using a movable mirror to reflect the body. At least two other bubble sextants were available commercially by 1914.[42]

The bubble sextant had come to stay; it would be the basis for celestial navigation of aircraft until gyro-stabilized, computer-directed photoelectric cells became available, to those who could afford them, after World War II. The bubble sextant did not drive other contenders from the field immediately, however; a number of pendulum sextants for air use appeared before 1914.[43]

Any navigational device is invariably judged by the accuracy of its results, and the early bubble sextants produced surprising accuracy. The results of two flights, made particularly to evaluate sextants, were published in detail. In both cases, a second observer noted the actual positions of the balloon, by map reading, for comparison with the celestial results. In May 1905 the Butenschön sextant was used on a day flight from Berlin to Upper Silesia. Seven positions were computed from observations of the sun and moon. The greatest error was 21 miles; the average was 10. In July 1908 the Ponthus et Therrode was used to determine two fixes at night. One was 6 miles in error, the other 5 1/2.[44] Such results were certainly adequate to warn balloonists that they were approaching the seacoast. One might expect that balloonists rushed to take up celestial navigation, but this was not the case.

The two test flights referred to above were made by men who had given long study to the problems of balloon navigation. They had much experience manipulating instruments generally and sextants particularly. They found that extreme care had to be used in measuring the heights of celestial bodies. Whenever possible, they took several observations and averaged the results. Crew coordination was important; the best observation would be ruined by another crew member's movement. Balloonists who were unable or unwilling to accept such discipline could not get satisfactory results with any artificial horizon sextant.[45] There was one other reason for most balloonists to avoid celestial navigation: measuring the height of the star was not the whole problem. The height was of no value until it had been reduced to a position on the chart.

The marine navigator reduced each observation by solving the celestial triangle (Figure 5) mathematically. He worked out the rather forbidding equations that expressed the angular relationships between the elements of the triangle—the observer, the geographic pole, and the point on the earth directly beneath the star—to 1/10 of a mile. With the requisite training, tables, and time, he reduced each sight by adding and subtracting logarithms. But "working out logarithmic computations is not everybody's thing (*nicht jedermanns Sache ist*)." No matter how mathematically inclined the balloonist might have been, working conditions in the basket were not conducive to such a procedure. From the beginning, the general use of celestial navigation in the air has depended on finding simpler and quicker methods of reducing the observed height to a position line on the chart. Simplification usually sacrifices precision, but the balloonist, drifting along over a solid layer of clouds and thinking about the ocean, would be happy to know his position within 10 miles or so.[46]

The simplest way to visualize the celestial line of position, and therefore a demonstration always found in the introductory celestial lessons in navigation schools, is to use a globe. If a star is at a height of 90° above the horizon, it is directly overhead. The position of such a star in celestial coordinates, given by an almanac, when converted to geographic coordinates is the position of the observer. If the star is, say, 40° above the horizon, the observer is 50° (90°−40°) from the star. With a globe, it is easy enough to plot the position of the star and swing a circle, with radius of 50° or 3000 nm (50° × 60 nm/degree), around that position. The circle is the line of position. It includes all possible points where the star will appear 40° above the horizon, and the observer must be on the line. The accuracy attainable is directly proportional to the size of the globe. One of the first balloonists to experiment with celestial navigation actually used a 10-inch diameter globe to reduce his observations. He claimed that he could work to an accuracy of 1/5 mm, or 6 nm,[47] but he must have used a very sharp pencil. Regardless of the size, a globe is not particularly handy to carry around, and it has since found little application in celestial navigation outside the classroom.

One might assume that a circular line of position could be constructed on a chart as easily as a globe. Unfortunately, this is not the case. The surface of a sphere can be represented on a plane only by accepting distortion. For the distances involved with celestial lines of position, the distortion becomes extreme. The simple method of plotting the star's position and drawing a circle around it cannot be used. The balloonist who would use celestial navigation had, therefore, to devise different methods.

The intercept (St. Hilaire) method of plotting the line of position (see above, p. 13) was applied to balloon navigation as early as 1903. St. Hilaire had devised a procedure that allowed almost all of the computation to be done before the observation, thereby reducing the delay between observing the body and plotting the position. He accomplished this by assuming a position close to the ship's actual position and working out the height and azimuth of (direction to) the body as they would appear from the assumed position. After the observation, the navigator needed only to find the intercept by comparing the observed height with the computed height. From the assumed position, he then marked off the intercept distance toward or away from the body and constructed the line of position perpendicular to the azimuth. He could easily plot the position line on the chart within a minute after putting down the sextant. For the marine navigator, the advantage of the method was speed. For the balloonist, an even greater advantage was that all the logarithmic computations could be done on the ground.[48]

With little knowledge of upper-air winds, the balloonist could not predict where he would be a few hours after he left the ground. He could not therefore determine assumed positions for precomputing celestial data that would be very close to the balloon's actual positions. In practice, therefore, the height and the azimuth of the bodies to be used during the flight were computed as they would appear at the point of departure throughout the expected duration of the flight. The intercept method was based on the assumption that the circular position line could be approximated by a straight line. For short distances the assumption was valid, but as the balloon moved farther away from the departure point, the intercepts became larger, and the straight line approximated the circular line less and less. Unfortunately, this source of error increased the farther the flight continued—precisely as navigation became more critical. There was another drawback to the use of precomputed data: a flight could so easily be postponed, and all the logarithmic computations rendered worthless.[49]

The next method devised to adapt celestial navigation to aerial use overcame both these objections. The 2nd International Congress of Aeronautics, which met in 1900, formed a subcommission to study the problem of finding one's position in a balloon. The subcommission briefly considered the question of aeronautical charts (see above, p. 19), but it gave most of its attention to the problem of celestial navigation. The most active member of the subcommission was L. Favé, the chief hydrographic engineer of the French navy. He developed a pendulum sextant, which was never widely used, but his attempt to produce a better method of sight reduction was more significant.

Favé recognized that precomputation was both essential and tedious. He devised a system in which precomputation for flights over most of Europe could be done "in advance and once for all." The Congress lacked the money to carry out the project, but Favé's work was practical, as later adaptations would show.[50] He described in detail all that needed to be done. Choosing a position in the Po Valley that he felt was central for European ballooning (could he have been influenced by the fact that the Congress to which he submitted his report was meeting in Milan?), he computed the height and azimuth for a particular star for every ten minutes of time. He then devised a table to show these values for 24 hours on a single sheet of paper. We commonly refer to the stars as "fixed," and as they appear from the earth, their declination (latitude) does remain almost constant. The apparent motion of the stars in longitude from one night to the next, caused by the earth's revolving around the sun, is regular, and a watch can be adjusted to run on sidereal (star) time.* On one sheet of paper, Favé therefore tabulated the height and azimuth for a single star, in sidereal time, as seen from a given position, and the data remained valid for about five years.[51]

The sun, moon, and planets, unlike the fixed stars, change constantly in declination. Height and azimuth of these bodies could not be tabulated so simply, but Favé demonstrated that a large graph could be constructed to simplify computations for them also. By determining the geographic location of any one of the bodies at a particular time, a balloonist could extract the height and azimuth from the graph without any arithmetical computations.[52]

For plotting the position line while in the air, Favé provided two charts, drawn on a common projection and scale, to be used one above the other. The top chart was a transparent map of Europe, centered on the Po Valley position that he had used in the precomputations. The other, of opaque paper, carried a set of arcs of concentric circles of varying radii, to serve as position lines. To trace a position line on the transparent chart, the balloonist had to align the set of arcs so that the height and azimuth of the particular body as seen from the center of the chart agreed with the height and azimuth precomputed for that location. After measuring the height of the body, the balloonist located the corresponding position line by visual interpolation and

---

* Time is an expression of the angular distance between a particular meridian and one of the celestial bodies. The most commonly used time is zone time, which is the distance between the sun and the meridian in the center of the time zone. In navigation, it is often convenient to use Greenwich mean time, the distance between the sun and the prime or zero meridian. It is equally feasible to define time with reference to the fixed stars. Sidereal time is measured from an imaginary point among the stars, the first point of Aries, which is the point where the sun crosses the equator northward in the spring. Local sidereal time expresses the angular distance between the first point of Aries and the observer's meridian.

traced it onto the top chart.⁵³ Favé thus provided circular position lines, avoiding the error of straight position lines with very long intercepts. He also showed that all of the precomputation could be worked by logarithms once, published as graphs and charts, and extracted easily thereafter as needed. Money being lacking, the project was not completed.

Favé's ideas were the basis of two plotting machines that were soon offered commercially in Germany. The first won a prize at the Frankfurt a. M. International Aeronautical Exposition in 1909. In this device, greater accuracy was possible because the charts were drawn to a larger scale; but they also covered a smaller area than Favé's. The position lines were provided on a long strip of linen that was mounted on two rollers, one on either side of the navigation chart. The latter was held firmly in a circular mounting. Both changes made possible more precise alignment of the two charts. The second model, available the next year, was more mechanized yet. It was a simple analog computer that used a thin strip of flexible metal to represent the position lines. The metal strip was positioned above the chart through a gear train. By applying varying pressure to the metal strip, it was bent to represent an arc of any desired radius. Both of these instruments were expensive, the first costing slightly more than the best air sextant, the second twice as much as the same sextant. Their use was therefore limited.⁵⁴

Several other methods of sight reduction appeared in the years before the war. Favé had described his system in 1906. The next year one of the French crews in the second Gordon-Bennett balloon race, flown from St. Louis, Missouri, devised a different graphical method. For latitude, they intended to use the simple method of latitude by Polaris (described above, p. 14). For longitude, they would need to use four different stars at various times during the night of the race. For each star, they precomputed a graph. Entering with the latitude (found from Polaris) and the observed height of the longitude star, the balloonists could find local sidereal time. This value, compared to Greenwich sidereal time on a watch carried for that purpose, was easily convertible to longitude. The graph did not receive the ultimate test because the weather was clear the night of the race, and the crew relied on map reading.⁵⁵ The idea was sound, however, and a variation of it was to reappear in the late 1920s as "Star Curves."

A third graphical method, proposed in 1912 for the trans-Atlantic airship flight then being discussed, used a round chart on the orthographic projection. By plotting the position of the celestial body on the edge of the circular chart, all position lines appeared as straight lines.⁵⁶ This obvious advantage was more than offset by the extreme difficulty of measuring distance on such a projection.

Inexpensive devices called "Transformators" were offered for solving spherical triangles. This device was composed of two circular charts, rotatable one over the other. A spherical triangle was drawn on the transparent top chart; then the chart was turned until one side of the triangle lay along the equator of the lower chart. At the equator, 1' of arc equals 1 nm in any direction, and the length of the triangle's side could be read directly in nautical miles.⁵⁷

A number of tables were published for balloonists who preferred to avoid graphs and diagrams. The accuracy of the particular table depended on the amount of interpolation required; to reduce interpolation, very large tables had to be carried. The number of options offered to the balloonist and airship pilot was so large that the editor of the reputable *Moedebecks Taschenbuch* of 1911 begged the question: "Looking back over the literature produced in the last few years on the subject of position determining, and with experience as to the usefulness of only a few of the particular methods available, it is scarcely possible to give a critical and impartial presentation of them here." He excused himself by appending a four-page bibliography of recent articles.⁵⁸ Had he been writing a few years later, he could have listed perhaps twice as many.

The multiplicity of methods resulted at least partially from a situation that still exists: some navigators prefer to work from tables, others from graphs, still others find mathematical methods most suitable. The manufacturers of navigational implements have had to cater to the whims of their clients ever since.⁵⁹ It also seems safe to conclude, however, that no single method was overwhelmingly superior to the others, so that the question remained open. While it has been possible to indicate that some instruments and methods were of dubious value, it is impossible to be precise in evaluating the usefulness of the remainder. The variables are too numerous, the effect of personal preference, training, and interest too great. What the literature does show is that the use of celestial navigation in balloons and airships was not widespread before the World War.⁶⁰ The basic problem had, nevertheless, been solved. Those who could afford to buy the instruments and would take the effort to master the techniques could find their position when above clouds or over water, usually to an accuracy of 10 miles or less.

For navigation over well-charted lands when the surface was visible, balloonists of course usually navigated by map reading. Above the clouds, celestial navigation was possible. For a time, balloonists hoped to make use of the earth's magnetic field for navigation when neither earth nor sky was visible. Many of the pioneers of celestial navigation also experimented with magnetic positioning.

The earth's magnetic field offers two variables that can be measured on the ground: dip and variation. The magnetic field is parallel to the surface of the earth only at the equator. At the poles, the field is vertical. In mid-latitudes, the field makes an intermediate angle with the horizon. This phenomenon, called dip, had been studied at length in perfecting the marine compass, and the value of the dip angle had been carefully measured in many places. Lines connecting points of equal dip run roughly east-west in Europe. If dip could be measured in the air, it would yield a position line that approximated latitude.

An instrument to measure dip was being used experimentally in balloons by 1898. In 1909 the Carl Bamberg firm in Berlin offered an instrument called the Double Compass for the same purpose. The measurement was difficult to perform, however, because the horizontal had to be defined as accurately as for celestial observations. The instrument was also susceptible to deviation. When balloons were supplanted by airships and airplanes, the greater use of iron, as well as electrical circuits, made the measurement of dip impractical.[61]

The other magnetic variable that might be used was variation, the angle between true and magnetic north. This value also had been determined accurately over much of the earth during the 19th century. Lines connecting points of equal magnetic variation were included on standard navigation charts, and in Europe the lines run roughly north-south. If variation could be measured in the air, it would yield a position line that approximated longitude. Theoretically, this could be done by taking a visual bearing of the sun or other body, using a bearing compass. The result would be a magnetic bearing. By using the best known position, the true bearing of the body could be computed. The difference between the observed magnetic bearing and the computed true bearing should then be variation; but in practice, there were complications. If the best known position was in error, the computed azimuth of the body would be in error also. The procedure assumed that there was no deviation in the compass, but compasses could rarely be perfectly compensated.[62] The attempt to determine position by measuring variation was abandoned before World War I. In modern practice, the problem is worked in reverse, to check the accuracy of the magnetic compass.

One other possibility for balloon navigation must be mentioned. The balloon drifts with the wind. If the direction and force of the winds at various altitudes could be known in advance, then the navigation problem could be solved before leaving the ground. Meteorology, however, has grown up alongside aeronautics, and very little was known about the upper air during the ballooning era. Indeed, it was long hoped that upper-air winds would prove to be as regular as the ocean currents.[63]

Methodical observations of the upper-air winds began in the 1880s at three locations in the United States, France, and Germany. The information gathered at the United States site, Blue Hill Observatory, Hyde Park, Massachusetts, was published for the benefit of aviators in 1911. The U.S. Weather Bureau started regular upper-air observations at all of its principal stations only in 1907. In the face of contrary data slowly accumulated by the meteorologists, and more rapidly by the aviators themselves in World War I, the hope of finding regular currents in the atmosphere gave way to a second: that "meteorological information will in time reach such a state that the conditions which the aviator will be likely to meet can be predicted quite accurately in advance."[64] The weather forecasters are still working on that one.

NAVIGATION OF AIRSHIPS AND AIRPLANES BEFORE WORLD WAR I

For many purposes airships are grouped with balloons as lighter-than-air craft. For nagivation, however, airships and airplanes have one feature in common that outweighs their differences. They are both powered vehicles. They can be steered. For navigation, each of them has a *heading,* which the balloon does not. The balloonist's approach to navigation might be described as passive. He sought to find where the wind had blown him, or the direction and velocity that the wind imparted to the balloon. Navigators would continue to be concerned with the wind, the motion of the air relative to the ground. In powered aircraft, however, they also have to deal with the vehicle's own motion relative to the air. The path of the aircraft over the ground is the resultant of two independent forces acting simultaneously. For this problem, neither balloon nor marine navigation offered a useful solution.

Marine navigators continually dealt with simultaneous forces. They resolved each force affecting the ship's position (ship's own movement, current, tides, leeway caused by a side wind) into N-S and E-W components, added them together, and found the net movement of the ship for a period of time (see above, p. 10). The method was mathematically precise, as with all marine navigation, but it was too involved for use in the air.

Because the first airships had such low speeds, less than 10 knots, the first discussions of the behavior of powered aircraft in a wind were concerned with finding the possible directions that an airship could move when the velocity of the wind exceeded that of the airship. In nautical terms, how close to the wind could the airship sail? Mathematicians attacked this question with trigonometric functions and constructed precise curves for various combinations of airship and wind speeds and directions.[1] By the time the problem of simultaneous forces had to be solved on board airships and airplanes, the theorists had provided a much quicker method—the vector diagram.

A vector is a line having direction and magnitude, used to represent a force. By drawing a vector to represent each of the two forces, that of the aircraft and of the wind, with the origin of the wind vector connected to the end of the aircraft vector, completing the triangle solves for the resultant force, the groundspeed and track (Figure 8). The vectors are drawn for a convenient time interval,

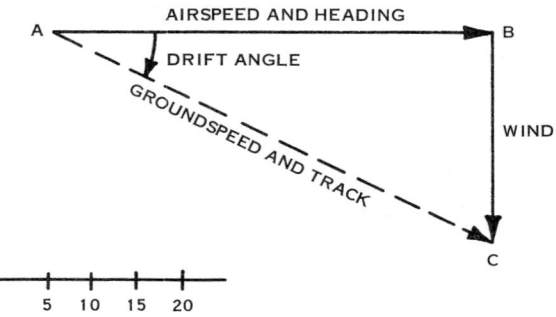

FIG. 8. The wind triangle. An aircraft steers east from point A, with an airspeed of 50 knots. The wind is blowing from the north with a velocity of 25 knots. The aircraft will never reach point B but will drift to its right. After an hour's time, it will arrive at point C, having traveled 56 nm over the ground.

usually one hour. If accurately constructed on graph paper, the track and groundspeed can be measured directly. As in any triangle, if any four parts are known, the remaining two may be found. This procedure was discussed in popular aviation magazines as early as 1910. Later the same year a simple analog computer to solve the same problem was described. It consisted of three rulers so connected that the triangle could be physically reproduced and the unknown values read from scales. The instrument, called a "compass for aerial navigation," was apparently the first of a long line of dead-reckoning computers for aerial use.[2] The first airplane pilots used neither graphs nor computers but approached the same problem empirically.

To fly to a visible landmark some distance away, the pilot simply pointed the nose of his airplane at the object. If there was a wind from the side, the nose of the airplane was soon pointing to one side of the landmark. The pilot turned back toward the landmark, but the airplane drifted off again. The more frequently the pilot corrected his heading, the more exactly the path of the aircraft over the ground approached that smooth curve shown in Figure 9a. The French call such a line a *courbe du chien,* the curve described by a dog running to catch his master. The dog does not consider his master's movement and consequently must be continually changing direction.[3] The fighter pilot knows the same line as a curve of pursuit. It is an inefficient course to follow, if the only purpose is to reach a given point, because it covers more ground than the straight-line distance. Solution of the wind triangle at the outset would give the heading to fly to compensate for the wind. Then, although the aircraft would never be pointed directly toward its destination, its ground path would be the straight line to that destination (Figure 9b).

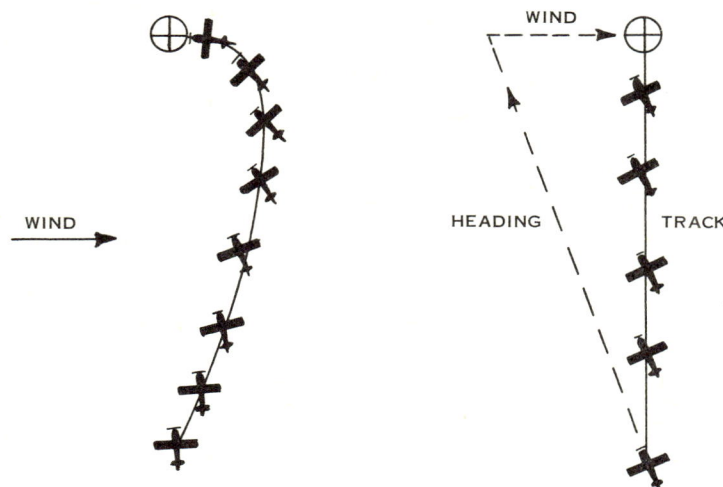

Fig. 9. Two ways to steer an aircraft to a visible point. a. Steering so that the aircraft continually points toward destination. b. By solving the wind triangle, the drift correction is found at the outset. The aircraft follows a straight line to destination.

Lacking an accurate measurement of the wind, the early pilot solved the wind triangle by trial and error. By first steering toward the visible landmark and noting the direction in which the nose of the airplane drifted off the mark, he learned not merely to turn back so that the nose of the airplane again pointed at the mark, but to turn further into the wind. Noting the position of the landmark frequently, he attempted to keep it at the same relative bearing as seen from the cockpit (e.g., in line with a particular cylinder head, strut, or rivet). He had then "killed the drift" and the airplane's ground path would be along the straight line to the landmark. This method served admirably for short, overland flights in fair weather.[4] As airships and airplanes became capable of flying a significant distance from the airfield, more sophisticated methods of navigation were needed.

In October 1906 Count Zeppelin's third airship succeeded in flying completely around Lake Constance. In 1908 his fourth flew 250 miles in 11 hours. In 1910 an offshoot of the Zeppelin company was formed to provide commercial airship transportation. Airplane development was not far behind. In July 1909 Louis Blériot flew across the English Channel. For ten minutes of the 37-minute flight, he told a reporter, he had been lost. "It is a strange position to be alone, unguided, without compass, in the air over the middle of the Channel. . . . I let the airplane take its own course."[5] In the summer of 1911, a number of cross-country airplane races were held in Europe. In each of them, the race did not necessarily go to the swift: the pilot's ability to follow the prescribed course figured largely in the result. By about 1908 to 1910, then, the navigation of airships and airplanes required practical attention.

Much that had been done in balloons was of course applicable in powered aircraft. Before the design of an aeronautical chart for balloonists had been settled, the chart makers were considering the needs of airships and airplanes. Groundspeed meters and the aneroid barometer were usable with slight modification. The airplane was, in general, a more rigorous environment for precision instruments than the balloon had been. To the extreme changes of pressure and temperature found in balloon flight, the airplane added severe acceleration, vibration, and changes in attitude. As an environment for man, the car of the airship was preferable to the basket of the balloon, but the early airplane was regressive. At first the pilot sat on top of the wing, exposed to the airstream. Even after he had moved into the open cockpit, he had to cope with cramped quarters and poor visibility of the heavens and the earth. Nevertheless, the instruments and procedures used in the navigation of balloons were applicable, with more or less modification, to the navigation of powered aircraft, and these modifications will be considered individually. Balloon experience had produced no instruments to measure heading and airspeed, however. As these instruments were to become of primary importance in airships and airplanes, they will be considered first.

For the essential purpose of determining direction in the air, some theorists suggested that aviators would require a homing sense such as that possessed by migratory birds and pigeons.[6] In the event, however, pilots had to make do with the more prosaic magnetic compass. Balloonists had borrowed the bearing compass from marine navigation. The pilots of airships and airplanes borrowed the steering compass from the same source. The steering compass lacks the sighting vanes of the bearing compass but has a lubber line, aligned with the fore-and-aft axis of the craft, against which the heading of the craft can be read. The marine compass could be used in airships with little alteration. For airplanes, however, extensive modification was required. As might be expected from their naval heritage, the British led the way.

In 1909, on one of the earliest flights made in England, S. F. Cody carried a marine compass supplied by the Admiralty.[7] The following winter a compass was displayed at the Aeronautical Exposition in Paris. In March 1911 three companies exhibited aeronautical compasses at the 3d Olympia Air Show in London. That same year an aeronautical compass appeared in the United States,[8] and the British Admiralty "officially designated an aeroplane compass" for the first time.[9] All of these instruments were slightly modified marine compasses. Pilots experimented with them because the utility of the compass was obvious, but the pilots roundly criticized the early models as untrustworthy.

Sailors quickly came to the defense of their instrument. The aviators, the sailors claimed, simply did not know how to care for or use the compass. Early aviators knew little of variation, less of deviation, and would express surprise "when a compass placed within 6 [inches] of the end of a stout movable iron bar does not point correctly!" On ships, compasses were kept as far as possible from all iron, and they were never mounted within 40 feet of motors. If this were observed absolutely in airplanes, "magnetic compasses would have to be discarded altogether."[10] Nevertheless, airplane designers could do much to improve the operation of the compass by being aware of the cause of deviation. Iron could be replaced with nonmagnetic metals in the cockpit. Movable objects, such as the control column, rudder pedals, and throttle linkage, absolutely had to be of nonmagnetic material. Otherwise, if a compass had been compensated perfectly, as soon as the rudder pedal was moved the deviation in the compass would vary.[11]

Designers also had to consider the location of the compass within the cockpit. The marine compass used a horizontal card. The pilot had to place his eye directly above the center of the compass rose to read it accurately; otherwise, parallax would introduce error. Further, "an aviator who has constantly to put his head to one side to observe his compass never looks at it sufficiently to steer a correct course." Compasses were usually mounted on the floor, between the pilot's legs, or sometimes on brackets from the side of the cockpit. The addition of a mirror, placed at a 45° angle above the compass rose, allowed the compass to be mounted on the dashboard in front of the pilot, at eye level. This was by far a more convenient position for reading the instrument. Unfortunately, the compass then added to the hazard of crash landings, because the pilot's head was too often thrown against the dash.[12]

Even when the compass was conveniently placed, the airplane pilot had difficulty reading it, because the British instruments were still marked with the traditional 32 points, plus halves and quarters (Figure 3). An airplane pilot could not afford to give his full attention to the compass, as the helmsman on a naval vessel did, and the old rose was not designed to be interpreted at a glance. A designer of British compasses commented: "My experience is that if you tell an aviator to steer say N. by W. 1/2 W., within two or three minutes he may be steering N.N.W. 1/2 W., or in fact any course but the right one, simply because, as he explains, he has forgotten how many 'black spots' of W. there were. A black division more or less means very little to him when his attention is fully taken up with other matters." Two corrective measures were indicated, and both were in general use by World War I. First, the 32-point compass rose was discarded in favor of the 360° graduation. Second, a movable pointer was added to the compass card, so that the pilot could mark his desired course on the compass itself. Then, if the movable pointer was opposite the lubber line, the pilot knew he was on course without reading

Checking compass deviation on an AT-7 navigation trainer in World War II. "Swinging the compass" had not changed very much. The tail is jacked up and the engines are running, to simulate flight conditions.
USAF photo.

either points or degrees. Some of the course indicators were quite elaborate. In one, colored lights warned the pilot that he was off course. In another, holes were drilled in the edge of the bezel for the insertion of tiny colored flags. The pilot could plan his flight on the ground and insert the number 1 (red) flag opposite the first heading he intended to fly, the number 2 (white) opposite the second, etc.[13]

The mount used with the compass was also critical. The marine compass was mounted in gimbals. When tried in an airplane, the gimbals transmitted the vibration of the engine to the liquid in the compass bowl, so that the card never came to rest. For a time, the compass bowl was simply nestled in horsehair or cotton, a temporary solution at best. By 1911 the Admiralty provided a shock-absorbing mechanical suspension for its first official airplane compass. Reversing the pivot on which the card turned further insulated the compass from engine vibration; and providing more space between the card and the edge of the bowl reduced the liquid swirl, which tended to start the card oscillating. Instrument designers also learned that the period of the compass must not equal the period of any of the oscillatory motions of the aircraft on which it was installed, because the needle would then swing wildly, even in level flight.* Thereafter, the liquid compass was reasonably stable in straight flight through calm air.[14]

Once magnetic materials had been eliminated from cockpit construction and improved mounts were in use, compasses could be swung for deviation and compensated using a procedure devised long before for ships. The airplane was placed in an open area, with the engine running, and the compass reading was checked against the reading of a master compass some distance away. The airplane was then turned a few degrees and the procedure repeated. After determining the error on each heading, small magnets were inserted beneath the

---

* The period of oscillation, which depends on the physical characteristics of a device, is the time required for one complete cycle; for an oscillating compass needle, it is the time from one extreme right deflection to the next. Several oscillatory motions occur in airplanes all the time. The tail moves right and left in a regular motion; the wing tips rise and fall. The engine sets up a number of complex oscillatory vibrations. The period of the early airplane compass happened accidentally to equal the period of one of the motions of the airplane. In this condition, called resonance, the weaker of the two forces absorbs energy from the stronger, and the results can be spectacular. Troops marching across a bridge are told to break step to avoid resonance. In the resonating compass, the needle swings farther and farther, until it finally achieves continual rotation. The problem of resonance also applies to the design of instruments incorporating air bubbles—sextants and bombsights—because the bubble has a period of oscillation in its chamber. Harry Egerton Wimperis, "Air Navigation," pp. 42, 44; Wimperis, *A Primer of Air Navigation,* pp. 24-27.

compass, until as much of the error as possible had been removed by trial and error. Although an airplane could be turned through the series of headings much easier than a large ship, the procedure was still longer than some pilots cared for. Others learned the importance of the instrument on cross-country flights, and they began asking to have their compasses swung before long flights, or after repairs had been made to their aircraft. The theory of compensation was widely known by the eve of the war, although in practice the well-compensated compass seems to have been the exception.[15]

Even after compensation, one major fault remained in the aeronautical compass: its behavior during turns was erratic. Sometimes the compass needle lagged behind the actual heading of the airplane; at other times it moved more rapidly than the heading was changing. After completing any turn, the pilot had to hold the airplane steady for several seconds to give the compass time to settle down on the new heading. This shortcoming of the compass was most critical when the early aviator found himself, by accident or design, inside a cloud.

Lacking any attitude reference under such conditions, the pilot tried to use the compass to keep the airplane flying straight ahead. The instrument was woefully inadequate for the purpose, but for some time the real problem was not even understood. "Some pilots claimed to have been in strange 'magnetic' clouds which caused the compass needle to rotate." Jean Conneau, who won three of the major airplane races in 1911, described his first attempt to fly through clouds: "Consulting my compass, I made straight north, yet I could not make out if I was following a straight line. I was caught in a whirlwind which made me face south, and was unable to understand this change of direction as I had not touched the rudder. I wondered if my compass was going mad . . . ." As late as 1915, a pilot report describing a similar experience ended: "I steered by my compass (which had recovered, being out of the clouds)."[16] Actually, the pilot could not hold his airplane in a level attitude lacking a visible horizon, and after the airplane started turning, the compass in due time noted the fact. If the pilot attempted to stop the turn by turning in the opposite direction, he almost always turned too far. Thereafter the pilot chased the compass needle, making larger and larger corrections, and the airplane usually fell into a spin. After falling out of the clouds, if there was enough altitude remaining, the pilot had a chance to recover level flight.[17] Pilots did not understand what was happening and blamed the compass.

For assistance on this problem, as well as for a general investigation of compass behavior, the Superintendent of the Royal Aircraft Factory, Farnborough, in late 1913 requested that the British Advisory

Committee for Aeronautics* study the instrument. The Committee's experiments, extending into the first years of the war, led to both an improved compass and an understanding of compass limitations. In particular, the strange behavior of the compass during turns was explained, and the need for a completely different instrument to indicate turns when flying inside clouds was recognized.[18] In 1914, however, airplane pilots lacked any reliable reference for steering inside clouds, and the unpleasant experiences with the magnetic compass discredited it in the eyes of most pilots.

In the years before the war, the second instrument needed to measure the aircraft's motion relative to the air, the airspeed meter, followed a course of development similar to that of the magnetic compass. The navigation of balloons, requiring neither instrument, provided no experience for the development of either. The problem of measuring airspeed, like that of measuring direction, was solved more easily for airships than for airplanes. Airplane pilots tried to use the compass as an attitude instrument as well as for navigation; they wanted an airspeed meter much more for the safe operation of their machines than for navigational purposes. And when the war started, few pilots trusted either instrument.

The airspeed of the first airships was determined by flying them back and forth across a measured course several times and averaging the elapsed times. At low wind velocities, the wind would aid the airship in one direction about as much as it hindered in the other.[19] The method was copied from that used in the trials of steamships, and it is still used today to calibrate airspeed indicators. On the first airships, the method sufficed, for speeds were low and engines simple; they were usually running at full power or not at all. As speeds increased and power plants became more flexible, however, a more sophisticated means was needed to measure airspeed.

Apparently the first instrument to be used was an air log, designed and named in direct analogy to the marine log. The air log was a hydrogen-filled balloon on the end of 100 meters of silk thread. The balloon was thrown overboard, and the time required for the string to be pulled off its reel was converted to speed.[20] This primitive method was better than nothing for airships, but it had no application

---

* The Committee was formed in 1909 to coordinate the aeronautical work being done in the National Physical Laboratory, the Army, and the Navy. A representative of H. M. Balloon Factory, South Farnborough, the predecessor of the Royal Aircraft Factory (later to be renamed the Royal Aircraft Establishment), was soon added to the Committee. The U.S. Committee of the same name, formed in 1915, was patterned consciously on the British model.

in airplanes; and the airplane pilot had a pressing need to know his speed through the air.

The lift generated by an airplane wing depends on the density of the air and the speed with which the wing moves through the air. As speed decreases, a wing reaches a critical point at which it will no longer support the airplane, and the airplane stalls. At the upper end of the speed range there is another critical point, that of structural failure. For landing, the pilot slows the airplane as much as possible, yet he must not allow it to go below stall speed. The first airplane pilots of necessity flew by "feel"; they founded the seat-of-the-pants school. By 1910, however, one pilot was pointing out that "it is almost impossible to estimate the speed at which one is rushing through the air . . . . Good judgment is, of course, everything in flying, but speaking from some considerable experience I can say that good judgment in estimating speed is not easy to acquire, and it is for this reason that I take the view that some suitable signal alarm would be of real practical value on an aeroplane." He asked for the development of an acoustical device to warn the pilot when he was approaching the critical speeds.[21] By 1910, the year the German airship transportation company was formed, both the airplane and the airship required an automatic means for continuous measurement of airspeed, the one for safety, the other for navigation. A meteorologist's instrument promised to fill the need.

Measuring the airspeed of an aircraft is equivalent to measuring the speed of the air past the craft. The anemometer was therefore taken from the weather station and installed on aircraft to measure airspeed. At the beginning of the 20th century, three kinds of anemometer were in general use: the rotating cup or propeller, the pressure plate, and the pressure or pitot tube. All three were used on aircraft. Each of them had to be placed beyond the region of disturbed air near the propeller. Far out on the wing was the only acceptable location on a single-engine airplane. The data gathered by the instrument had either to be transmitted to an indicator in the cockpit, or else applied directly to an indicator large enough to be read from the cockpit.[22]

The revolving cup anemometer employed three or four hemispherical cups on the ends of short arms, rotating about a vertical shaft. The instrument actually measured the volume of air that moved past it, and the rotating shaft was geared to a counter. The meteorologist used a watch to find the velocity of the wind, by noting the number of revolutions made in a minute and then converting that value to meters per second or miles per hour. For use in an airplane, the conversion obviously had to be accomplished automatically.[23]

Mechanical conversion was simple enough—it is accomplished by the ordinary automobile speedometer—but transmitting the data accurately to the cockpit was almost insoluble. For use on an airship, this problem did not arise, for the instrument was mounted in the control car and could be read directly. The instrument was also well suited to airships because it measured low speeds more accurately than other anemometers. In 1914 the German *Albatros* biplane was equipped with a revolving cup airspeed meter; the dial was large enough to be read from the cockpit. The Germans used this instrument widely during the war.[24]

The second category of anemometer was the pressure plate. This simple device measured the force that the wind exerted against a flat plate held perpendicular to the airstream. The plate was hinged at one side and restrained by a spring. The wind pushed the plate out of the vertical, and a pointer attached to the plate moved over a scale indicating speed. This kind of anemometer was widely used by French meteorologists, and a French aircraft instrument operating on this principle appeared in 1913. The indicator was mounted on the wing, thus avoiding the problem of transmitting data to the cockpit. The French instrument bore only a single mark indicating the "normal" speed of the airplane. A similar Italian instrument, employed on the Caproni bombers of World War I, added two additional marks: maximum—the danger point in a dive, and minimum—the stall speed.[25]

The third kind of anemometer, the pressure or pitot tube, was ultimately to become the standard airspeed meter. Before it was adapted to aircraft, it had long been used in the laboratory as well as in the weather station. Its principle was simple: compare the pressure of the moving air, sampled by an open tube pointing into the airstream, with the static pressure, sampled by a tube protected from the airstream. The instrument compared the two pressures by applying them to opposite sides of a U-tube containing colored liquid. The liquid dropped on the impact side and rose on the static side, and the tube was calibrated to indicate speed. Under the controlled conditions of the laboratory, the pitot tube was a precision instrument. In an airplane, however, the liquid was affected by all accelerations and gave erroneous results when the airplane was climbing, descending, or turning.[26]

From the fall of 1913 through the spring of 1914, a spirited argument raged in the British aeronautical press over the merits of the various airspeed meters. The cup anemometer measured low speeds accurately. It was also little affected by changes of air density, giving a reasonably accurate measurement of airspeed at all altitudes aircraft could then reach. Because the lift of an airplane is dependent on both speed and air density, an instrument that responded to density changes seemed preferable for use in airplanes. Most of an airship's lift, however, came from the lighter-than-air gas. The cup anemometer was the obvious choice for the airship. The pressure plate was unaffected by airplane maneuvers, which was a desirable characteristic; but the plate, exposed perpendicularly to the airstream, increased the drag that the engine had to overcome, and the instrument could not easily be coupled to a remote indicator in the cockpit. The pitot tube was sensitive to variations in air density and therefore made a particularly good "buoyancy meter" for indicating stall speed. It did not add appreciably to the airplane's drag, its results were easily transmitted to the cockpit in lightweight, flexible tubing, but as long as liquid was used to compare the impact and static pressures in the indicator, it could be a dangerously misleading instrument.[27]

The way out of the dilemma was found in the spring of 1914 with the introduction of pitot instruments that used elastic diaphragms, instead of liquid, to compare the two air pressures. As in the aneroid barometer, a slight movement of the diaphragm had to be multiplied through a series of pivoted arms to secure a discernible movement of the pointer across the airspeed dial. Much patient work would be required to perfect such a precision instrument. Because the principle was first applied only on the eve of the war, many aircraft designers understandably continued to install the earlier airspeed meters. Most pilots were suspicious of the instruments, also understandably. One of the "immediately pressing problems" taken up by the U.S. National Advisory Committee for Aeronautics after its formation in April 1915 was "the devising of accurate, reliable, and durable air-speed meters."[28]

Pre-World War I pilots understood the relationship between the wind, the motion of the aircraft through the air, and the motion of the aircraft over the ground; but the meteorological organization to provide frequent measurements of upper-air winds was in its infancy, and the instruments to measure aircraft heading and airspeed were rudimentary. How, then, were they to navigate as journeys became longer? They had to continue to rely on visual observation of the earth. Map reading was to remain by far the most common navigational procedure. The instruments used by balloonists to measure groundspeed and drift (hereafter called driftmeters) also promised to be helpful.

The airship was a better platform for the driftmeter than the balloon, with its continual rotation, had been. The magnetic compass also worked reasonably well to measure heading of the airship. Drift from the driftmeter applied to the heading gave the airship's track across the ground. Groundspeed could be found from the driftmeter, subject to the same errors encountered in using the instrument in balloons: inaccurate knowledge of the height above the ground and inability to maintain the optics level with the horizon. The ingredients for dead reckoning were thus available to the airship navigator, but the inaccurate groundspeeds precluded accurate DR positions. Airship navigators therefore had to determine their positions from map reading at frequent intervals to cross-check the dead reckoning.[29]

The driftmeter was even less satisfactory in the early airplanes. The first of several driftmeters for airplanes, the Daloz compass, appeared in France in 1910. This instrument consisted of a magnetic compass with a transparent card on which was etched a set of parallel lines. The magnetic needle turned with the card, but the two could be disengaged by a clutch arrangement, which allowed the pilot to set the angle between needle and parallel lines at any value he chose. The compass was mounted in the floor of the cockpit, above a lens that focused a view of the earth onto the back of the card. To use the device, the pilot first measured the course he wanted to follow on a map and then drew a line in that direction on the ground at the airfield. He next moved the airplane over the line on the ground, disconnected the transparent card from the magnetic needle, turned the card so that the parallel lines were aligned with the line in the dirt, and then connected the needle and card again. In flight the pilot steered the airplane so that objects on the surface appeared to move along the parallel lines. Two additional lines, perpendicular to those used to keep the airplane on its course, were provided for measuring speed. The two speed lines, in conjunction with the lens, subtended an angle of 45°; therefore the distance on the earth's surface appearing between the lines equalled the airplane's height. The time required for an object to move from one speed line to the other could be converted to groundspeed. The inventor admitted that, lacking accurate knowledge of altitude, groundspeed so determined would be of slight value.[30]

Variations of the Daloz compass appeared in England and Germany in 1912 and 1913. Although the later instruments provided for setting the course in flight, without drawing lines on the airfield, they all suffered from common difficulties. The magnetic compass on which they were based was prone to all the shortcomings previously described. Early airplanes were notoriously unstable; they seemed to expend as much energy in pitching and rolling as in forward motion. The grid lines therefore wandered back and forth over the landscape. The combination of wandering magnetic needle and wandering grid lines produced a picture in constant flux. As so often happens, an attractive idea proved impractical. The Aeronautical Society of Great Britain was told in 1913 that progress was being made with the airspeed indicator and compass, but not in the development of instruments to measure track and groundspeed. The speaker closed on a glum note: "I see no satisfactory solution of this problem, and will not discuss the question, except to suggest that someone here present solves it during the coming year."[31] It took somewhat more than a year. Meanwhile, airplane pilots navigated by map reading, with some assistance from the compass.

Although the balloonists had gotten work started on aeronautical charts, not much had been accomplished when airships and airplanes started flying beyond sight of their airfields. Practitioners of the three different modes of aerial locomotion made much of their different cartographic requirements, but on this subject they really had more in common than they knew. All of them needed charts designed to show those landmarks that would be easily recognizable from above, charts that included essential aeronautical data while suppressing needless detail. Their principal difference actually concerned the size of the chart. The airship navigator had room to spread out the largest chart. The airplane pilot did not.

Using charts of any size while sitting on the wing of the first airplanes was indeed a challenge. The only way to retain a piece of paper under such circumstances was to anchor it securely and protect it from the airstream. The solution, which seems to have been adopted universally, was to cut charts into long strips, covering only a few miles either side of the intended course, paste successive strips together, and mount the finished product on two rollers. As the flight progressed, the chart was rolled from one roller onto the other.[32]

After the pilots moved into open cockpits, they continued to use maps on rollers, because they had no space to spare, and it was very easy for loose paper (or gloves or goggles) to be sucked overboard. Deluxe chart holders came with a celluloid cover to protect the map. Given the amount of oil thrown back by the engine, it was worth the extra cost. One could not write on the celluloid cover, however, and a competitive model was introduced. It stored two strips of paper to be unrolled simultaneously. The chart was beneath a strip of transparent paper. The chart was protected, and the pilot had a place to take notes and make computations. Most chart holders displayed 20 to 25 miles of the 1:200,000 scale chart at one time. At the speed of early airplanes, this distance was equivalent to 25 to 40 minutes of travel.[33]

As slow as the first airplanes were, they were much faster than balloons. Airships could fly very slowly or stop if, for example, the crew had difficulty finding a landmark. The view from the airplane was much more restricted than from either balloon or airship. Further, the airplane pilot could devote only a part of his attention to navigation. To facilitate their task, experienced airplane pilots agreed that considerable preparation, other than merely cutting and pasting the charts, was worthwhile. The course was drawn on the charts before they were cut to fit the rollers. Measuring distances on the chart was difficult, if not impossible, in the air. Pilots therefore divided the course into convenient increments, marking the distance remaining to destination at each division. The increment chosen depended on the scale of the chart and the airspeed of the airplane. The author of a German textbook recommended 3.75 cm, which on the General Staff chart was five minutes at 90 km per hour. This interval had an added advantage: it was "a full two fingers' width" for quick approximation in the air.[34] Another author encouraged study of the chart, to identify landmarks and hazards ahead of time. "As one reads the play before the theater performance for a better understanding, so one must study the chart. This is the best protection against surprises."[35] Because the rollers held the chart in a fixed position, north could not be kept at the top. If a turning point was planned for the flight, the course still had to go down the middle of the strip of chart; the direction of north would therefore change from one part of the chart to another. A French textbook recommended marking the direction of north at intervals along the course, as well as drawing in landing fields with elevations. "But do not encumber the chart with a multitude of remarks; this will make reading it very difficult. The chart is almost always complicated enough by itself without obscuring it further."[36]

Only three years before World War I, most airplane pilots had no option but to rely solely on map reading for navigation, for their aircraft lacked even the most rudimentary instruments. When "the mobilized military air force of the British Empire"—six pilots—took part in the August 1911 maneuvers in Cambridgeshire, their airplanes had "nothing but a revolution counter, and sometimes, though not always, a compass. If the pilot's scientific ambitions went beyond this simple outfit, he carried a watch on his wrist and an aneroid slung round his neck." For the first airplane races that same year, the machines were similarly unequipped. During the Saxony Race and the German Circuit, both held in the early summer, the weather was far from favorable. The races extended over several days, one stage ordinarily being flown each day. Most of the airfields were within or near large cities, and picking up the first landmark in the morning haze, with no directional reference other than the sun, proved diffi-

Anthony G. Fokker on one of his airplanes about 1911–12. He is wearing a wristwatch and has an altimeter strapped above his knee.
USAF photo.

cult. Lieutenant Hans Steffen, on the staff of the Noncommissioned Officers School, Potsdam, flew in both races. He recommended flying around each airfield, before the race, to choose a landmark—the higher the better—for the initial checkpoint. When the contestants ran into low clouds and fog along the route, they usually continued, hoping for a break in the weather, and often came to grief. A rather common error was taking the wrong fork when following a railroad.[37] After compasses were installed, it was easy to determine if a particular railroad lay in the general direction shown on the chart.

A magnetic compass added considerably to the assurance with which a pilot could find his way across country, even if the needle oscillated much of the time and the direction was no more than approximately correct. The degree of trust pilots placed in the compass varied widely. For years the fact that anyone made a long flight on compass heading alone was worthy of mention in the aeronautical press. Lieutenant Conneau, who won three of the 1911 races, was a midshipman on a French warship before he took extended leave to learn to fly. His naval background probably accounts for the keen sense of navigation that contributed to his success. Nevertheless, he believed that some of his British competitors were wasting their time in trying to compensate their compasses perfectly. "I told them that two or three degrees variation in a compass was not of much consequence, considering the deviations in flying."[38] When a pilot found his checkpoints covered by clouds, or when crossing the English

Channel, he had no choice but to fly by compass heading, if he had the instrument installed.

During such periods, it was useful to know the airplane's groundspeed, to help identify a landmark once the cloud cover, or the Channel, had been crossed. Careful pilots checked their groundspeed frequently between map-reading fixes. Then, when clouds intervened, they could compute their approximate position on the map. If a hole through the clouds appeared, they were better able to identify a precise landmark quickly, before the clouds closed in again. Pilots who took the trouble to compute groundspeed before they needed it, on the chance that they might, were practicing a defensive approach to navigation that has been too rare. The approach pays dividends, because the individual is better prepared for the unexpected. He also learns to trust his instruments and calculations before he is placed in a position where he has to do so. The price paid is a great amount of work that, to the casual observer, seems wasted.[39]

Navigation was difficult, particularly for the pilot flying alone. Early airplanes were unstable. Without constant pressure on the controls, the nose or a wing dropped, and the airplane was soon in an unsafe attitude, or at least going in the wrong direction. Under such conditions, the pilot could not give more than a fraction of his attention to navigation; and that often was not enough. Two remedies were proposed to aid the pilot. One group of people pointed to the ease of navigation on board ships and airships. They saw the solution in adding a second man to the airplane crew, charging him with navigational duties. The second group concentrated on making map reading easier, and they urged the erection of unambiguous, artificial landmarks all over the countryside.

Temporary landmarks were provided to help the pilots find their way during the early cross-country races. When Cody was attempting to win the *Daily Mail* prize for the first London-to-Manchester flight in 1909, he arranged for a number of navigational aids. At several places, he had white sheets 100 feet long spread out on the ground, with men standing by to wave large white flags. Each flag carried a number that identified its position. When the airplane came into sight, the signalman was alternately to wave the flag vigorously, then to point it in the direction of the next town. Smoke rockets and captive balloons were also used at critical points. Cody could not start his engine on the appointed day, and he turned his attention to other matters. When Louis Paulhan and Claude Grahame-White raced for the prize the next spring, they relied primarily on the London and North-Western Railway. The Railway actively cooperated. "In order to assist the aviators in picking their way past complicated crossings, the sleepers [crossties] on the right line to follow were whitewashed. This was extremely helpful . . . ."[40]

During the many races of 1911, various devices for marking air routes were tested. The European Circuit included a crossing of the Channel. The officials of the race had a large arrow chalked on the ground near Calais, pointing toward Dover, to help pilots establish the correct direction before losing sight of the French coast. The winner of the race generally disparaged the other expedients. Signals very close to the ground often could not be seen. Balloons were too often blown away. Fires were apt to be lighted and flags put on public buildings for purposes other than guiding aviators. He concluded that, in the future, permanent, white pylons would have to be erected along major air routes.[41]

Permanent markings of a different kind were already appearing. One of the first French airfields was at Pau, in southwestern France. In late 1909 the president of the local aero club wrote the mayors of the neighboring towns, asking their assistance in painting the names of the towns on the roofs of large buildings. The following year a plan was announced for marking key localities throughout France. Rather than the name of the town, however, the distance of the place north or south and east or west of Paris was to be shown. The design of the sign resulted from experimentation at the Eiffel Tower. With signs of different size and shape spread out on the ground, observers on the upper platform chose a 6-foot sign, made of silvered glass balls on a black background, as most suitable. In Belgium, a painted roof sign was favored: a white cross, with the name of the town painted on the cross arm, and smaller arms pointing to the landing field and hazards to flight. A German aviator, noting that church steeples often stuck up through fog banks, suggested numbering all the clocks on steeples and towers and displaying the number on the face of the clock.[42]

Many theorizers proposed elaborate grids for dividing up the countryside, based either on latitude and longitude, distance from the national capital, or the index to the sheets of the national aeronautical chart. Each scheme included a proposed ground mark to tell aviators that they were within a particular grid square. The more complicated schemes included methods to show the approximate position within the grid square. For any such scheme to be effective, a vast number of ground signs would have to be installed and maintained. To justify the expense, supporters argued that such signs were analogous to road signs and aids to marine navigation, both of which were supplied by governments for the use of their citizens. Roof signs appeared in a few places, but no country carried out any of the more comprehensive, and costly, schemes. The volume of air traffic did not warrant it. Not only would the initial cost be exorbitant; all signs would have to be repainted at intervals, and snow would have to be removed. Rather than "riddle the earth's surface with innumerable artificial landmarks," an early critic urged that aviators be taught how to navigate.[43]

Even the critics admitted that ground marks along borders might be worthwhile. By 1913 the European nations were prohibiting flights over areas of military importance and requiring that aircraft cross borders only at designated points or through specified corridors. Ground marks warning pilots of these areas also seemed necessary. But such marks were warnings against immediate hazards, as much as for orientation.[44] For flying at night, however, there was general agreement that a number of beacons had to be installed.

The Germans took the lead in developing beacons as aids to aerial navigation. In 1909 and 1910 experiments were conducted with lighted balloons, electricity being supplied by the line that anchored the balloon. Such lights were not visible very far, and attention was next directed toward the design of aerial lighthouses. At first it was widely assumed that the best beacon would be a searchlight pointed straight up. Experiments with a vertical beam were disappointing, because it could not be seen from very far away. A better design used a series of lenses in a circle around a fixed light source, each lens focused at an angle only slightly above the horizon. A supplementary lens directed a small amount of light upward, so that the beacon would be visible from directly overhead. For even greater range, a third model focused all of the available light into a single beam that was very narrow in the horizontal dimension, wider in the vertical. The beacon was then rotated through 360°. The use of different colors would have helped to identify individual beacons, but the introduction of a colored lens seriously reduced the light output. Therefore beacons remained white. For identification, the stationary beacon flashed a letter in Morse code. The rotating beacon could not be flashed, because an observer would not be able to see the entire sequence. A second light was sometimes installed nearby, to flash an identifying signal. Electricity was the preferred power whenever available. For use in remote areas, as well as for portable military beacons, the Germans also produced acetylene models. By the early summer of 1914, Germany had 21 large aerial beacons in use, mostly at airfields. France and England had none.[45]

Thus at the outbreak of World War I the need to simplify aerial navigation by providing ground signs and beacons had been widely discussed. Beacons were considered essential for night flight; ground marks for daytime use were desirable at least in certain cases. Nevertheless, few of either had been installed.

The other possibility for improving the navigation of airplanes was to add a second man to the crew. Arguments for this step included the analogy to ship and airship navigation, the slight attention a pilot could spare from control of the airplane, and the inability of a man flying alone to use the more complex navigational instruments.[46] One obvious cost of a two-man crew was shorter range. Early airplanes had little excess lifting capacity, and the addition of a man meant the removal of gasoline. Another problem was crew coordination, especially communication in flight.

In the 1911 races several of the contestants got lost. Flying without compasses, or with compasses of doubtful reliability, in marginal weather, at a time when no one knew much about cross-country flying, it was easy to make a case for carrying a special crew member to do the navigation. The winner of the Saxony Race was leading in the German Circuit, until he took the wrong fork at a railroad junction. Another pilot wandered through the Harz Mountains for 2 1/2 hours, finally crash-landing at twilight. Several pilots had to fly in circles for a while, until they could locate their positions. Lieutenant Hans Steffen, whose map-reading experiences during these races have been described earlier, was convinced that "for navigation, a passenger skilled in charts and compass belongs in the machine." Gustav Hamel, at the time of his death in 1914 England's most famous cross-country pilot, agreed: "Finding the way is greatly facilitated" with a second man along, "for the passenger has more leisure to examine the maps and observe the country than the pilot."[47]

Once beyond the map-reading stage of navigation, the case for a specialist crew member was even better. Some navigational devices the pilot simply could not manipulate. A driftmeter gave poor results when the pilot could only glance at the instrument occasionally. Better results were being achieved with more complicated driftmeters on airships, but "the driver of an aeroplane has all he can do to manage his machine, without peeping through little telescopes, twiddling mirrors, and reading scales."[48] Celestial navigation was clearly impossible in the single-place airplane. Almost nothing seems to have been done to adapt celestial navigation to use in airplanes of any size before the World War.[49] Looking ahead to the advent of the long-range airplane, Hamel foresaw the need for determining position with the artificial horizon sextant, "not by the man who is controlling the machine, of course, but by an assistant."[50]

The second crew member introduced a new requirement: the coordination of actions by individuals, the exchange of information and intentions. Speech was drowned by the noise of the motor, but notes could be passed back and forth. One airplane crew prepared identical maps, wrote comments and questions in the margins, and traded maps periodically. Hamel used a method that surely encouraged brevity: he turned off the engine, said what he had to say while gliding, and started the engine again. Primitive speaking tubes were employed but apparently were more trouble than they were worth. Perhaps the most satisfactory method for exchanging information was through hand signals. More than a means of intercommunication was required, however. Duties had to be divided, responsibilities established. The men had to learn to work together, toward a common goal, in an occupation that had been as individualistic as riding a horse. Being engaged in a hazardous pursuit, they had to learn to trust each other. That the latter was not automatically accomplished is indicated by the advice of a German officer to student aviators: "Nothing is more crushing and undermining of mutual trust, than for the observer to act as though he still knows the position, when he lost it several minutes before."[51]

The reader may have noticed a certain reluctance to name the new member of the crew. Lieutenant Steffen was perhaps the first to grapple directly with the problem. He called the pilot *Fahrer*, driver, and the new crewman *Führer*, leader or director. Steffen's being a *Führer* may have influenced his choice of words, but during

Curtiss flying boat *America*, built for 1914 Atlantic attempt. *USAF photo*.

World War I the Germans relied extensively on noncommissioned pilots to chauffeur commissioned officers who navigated, observed the battlefield, threw bombs overboard, and manned flexible machine guns.[52] The Germans adopted the word *Beobachter,* observer, for this aircrew position, and both the English and the Americans later used the term. The French used both *navigateur* and *observateur.* The terminology has continued to be troublesome, as crew duties have been divided and combined, navigational duties sometimes being shared by more than one individual, sometimes being accomplished by a crew member who spent much of his time on other duties. Whatever the navigator might be called, the utility of carrying him on long flights, or when navigation was apt to be difficult, was generally accepted by 1914.

Much too early in the history of powered flight, men began talking about flying the North Atlantic. Such a flight would apparently require the services of a specialist navigator. An airship attempted the crossing in 1910, and when the war began plans were well along for a flying boat to do so. The navigational techniques used and contemplated on these flights may be taken as indicative of the best available at the time. In each case the importance of navigation was explicitly recognized.

The airship *America,* commanded by Walter Wellman, set out from Atlantic City on 15 October 1910, bound for Europe. The airship had been designed in Paris for a flight from Spitsbergen to the North Pole, and abortive flights had been made in 1907 and 1909. When Perry reached the Pole, Wellman turned his attention to the North Atlantic crossing. He secured financial backing from three newspapers and redesigned and enlarged the airship for the longer voyage. Of the crew of six, Wellman and two others had the Spitsbergen experience behind them. Of the new men, one was Murray Simon, a navigator from the White Star Steamship Company. Another was a radio operator. Radio was used only for communication, however; directional wireless was still in the future. No test flights were made. Three days after the manufacturer reported the craft complete, the voyage began. The helmsman found that he had to cut a hole in the canvas in order to see forward. The larger of the two engines failed the first day out; on the fourth day, the crew abandoned ship when in sight of a steamship, 375 nm east of Cape Hatteras.[53]

The navigator, Simon, relied on dead reckoning, updated periodically by celestial observations, exactly as he was accustomed to do on board ship. For direction, a steering compass was installed in the control car, and a master compass high up among the gas bags, where it would be less subject to vibration and deviation. Drift was measured

Navigation of Airships and Airplanes

61

Interior of Fokker Tri-Motor used on the Hawaiian flight, June 1927.
USAF photo.

The Boeing Clipper over Seattle, November 1938. The astrodome is visible
behind the teardrop-shaped DF antenna.
USAF photo.

visually by comparing the angle between the wake left by the guide rope, which was in the water continually, and the fore-and-aft axis of the airship. Speed was measured by towing a patent log in the water. Simon intended to use a marine sextant, and he prepared a dip table to correct observations taken as high as 6,000 feet. He also carried an artificial horizon sextant of the pendulum variety, for use when the sea horizon was invisible. He entered in his log: "Took several observations and decided that, like the rest of these artificial horizon sextants, it was not a particle of good." Although the crew had been unwilling to specify where they expected to strike Europe ("Any spot between Gibraltar and the North Cape will look good to us"), Simon was able to determine his position with sufficient accuracy for the rescue ship to intercept the drifting airship.[54]

After this failure, the possibilities of trans-Atlantic flight continued to be discussed from time to time until the outbreak of the war, and the navigational requirements were not overlooked.[55] The activity did not progress beyond the discussion level, however, until Glen Curtiss built a flying boat in 1914 to attempt the crossing. He engaged Lieutenant J. C. Porte, R.N., as chief pilot and navigator. The flight was canceled on the outbreak of war, apparently before Porte had completed his navigational plans. He had requested the assistance of ships in the North Atlantic. They were to fly signal flags indicating their distance north or south of his published route. He was going to use a driftmeter to measure drift and groundspeed. Several imaginative devices for reducing celestial observations were being considered. One method called for a mirror to reflect a portion of the sky onto a map mounted on the ceiling of the airplane cabin, where the angular distance of a body from the zenith was to be measured. Another scheme was to photograph a celestial body and the horizon simultaneously, develop the plate, and measure the distance between the two. The *Scientific American*'s correspondent, reporting a visit to the Curtiss factory, added a footnote: "It is still an open question as to how nearly dead level an aeroplane can be held for a brief time, say half a minute, in smooth air."[56]

Porte did not get to try any of the new techniques at that time. On the outbreak of the European war, the North Atlantic flight was postponed. When it was taken up again in 1919, airframes and engines would have a much better chance of completing the demanding trip. The instruments and techniques of aerial navigation would also have been improved, though to a lesser degree; and aerial navigators would have gained many hours of experience, though on much shorter flights.

NAVIGATION IN WORLD WAR I

Military aviation was a negligible quantity, little more than a curiosity, before World War I. By November 1918 the British alone had flown more than a million hours.[1] For so much flying, there was little navigation. Most of the flying—battlefield reconnaissance, artillery spotting, air-to-air combat, bombing of troops and supplies in the battle zone, interception of bombers by home defense forces—involved only the most rudimentary navigation.

The first reconnaissance missions by the Royal Flying Corps were flown in difficult weather on 19 August 1914. Two pilots, operating out of Maubeuge, Belgium, intended to stay together for mutual support as far as Nivelles, 28 miles away. "The machines lost their way and lost each other." One pilot "flew over a large town which he failed to recognize as Brussels," followed the Sambre River back toward his base but missed it and landed 20 miles upstream. The second "attempted to steer by compass through the banks of clouds, and after two hours of wandering landed," asked questions, took off and got lost again.[2] In the last year of the war, a pilot of the Royal Air Force found himself outnumbered in a dogfight over enemy territory. Prudence dictated retreat. He had a compass on board, but: "It was then that the thought of 'wind and smoke' struck me. I knew that the wind was due West when I left the ground, so I took a last glance at the smoke on the ground and turned up the wind indicated by the way the smoke was blowing." He climbed into the clouds, eluded his pursuers, came down out of the clouds over the lines, recognized the countryside, and landed at his home base. In four years of war, most pilots never advanced beyond the map-reading stage of navigation, but they learned to wring the last iota of intelligence from what they could see on the surface of the earth.[3]

Almost complete reliance on map reading should be no surprise. Before the war, the only generally accepted role for the airplane in combat was reconnaissance.[4] If the pilot had to be able to see the surface of the earth to perform his assigned mission, he would also be able to see landmarks to guide his flight. While the number of combat roles increased during the war, the visual element remained universal. Artillery spotters had to see the target and the impact of the shells. Weapons, whether bullets, bombs, or steel darts, were delivered visually. Of course there were times when, although the target area was clear, clouds covered the home base or the route to the target; but airplane ranges were short. The same weather system was apt to encompass both base and target. If the weather deteriorated before a pilot could return to his base, he was expected to make the best of a bad situation. Night flying complicated the problem, but as night flying increased, artificial, lighted landmarks were installed, and navigation continued to be by reference to the ground.

The navigation training of prewar aviators did not go beyond the essentials of the map and the compass, and many pilots ignored the latter. The student pilot spent most of his time learning to fly the airplane. The lectures and practical work devoted to map reading for navigation were a minor part of the curriculum. In cross-country flying, the pilot was taught to be constantly aware of the terrain—but the purpose was to have an emergency landing place in sight, because engines failed frequently. With the approach of war, the British schools cut the length of the flying course in half, so as to double the output of pilots. During the war the continual effort to build up the air forces, while making good combat losses, militated against the inclusion of anything not absolutely essential in the training programs. Anyone could appreciate that better trained pilots would reduce combat losses, but to harassed commanders at the front, half-trained pilots seemed preferable to no pilots. When courses were lengthened, specialization in the tactics of aerial fighting, bombardment, or observation was more important than navigation. For most pilots, navigation continued to mean map and compass, and new pilots invariably arrived in the combat zone with little practical experience behind them.[5]

Observers, in Great Britain at least, received less navigational training than did the pilots. Early in the war, service squadrons trained their own observers; only in 1918 did the British open a separate school for the purpose. The subjects included the flexible machine gun, aerial photography, cooperation with surface forces, and radio for air-to-ground communication; navigation was included only incidentally. In Germany, however, where the commissioned observer was likely to do the navigating, pilots and observers received similar training in navigation.[6]

Instruction in the schools changed during the war, as combat experience dictated, to include the aspects of map reading that had been found particularly useful. The armies and navies used charts with overprinted grids of perpendicular lines for quick and accurate designation of positions. To cooperate with surface forces, both pilots and observers had to be proficient in the use of the grid. Pilots were taught the peculiarities of French charts and what French first, second, and third class roads looked like, on the chart and from the air. They were taught that bodies of water were excellent landmarks except after heavy rains; that following meandering British rivers took too much time; that railroads were excellent to follow, but that "railways sometimes seem to end abruptly, which means that you are looking at a tunnel." They were shown how to make the magnetic compass approximate the true direction, either by attaching small compensating magnets to the compass or by displacing the lubber line an amount equal to the local variation. Either shortcut worked, as long as the airplane stayed in a limited area where variation did not change appreciably. Elementary astronomy was mentioned, so that the aviator could find Polaris at night and use the sun for approximating direction by day.[7]

Such rough and ready methods sometimes were not enough. Too frequently pilots landed at the wrong airfield. At times an airplane being ferried from England to France landed on the wrong side of the front. The most embarrassing such incident occurred on 1 January 1917. One of the first twin-engine Handley Page night bombers landed in a field and was surrounded by German soldiers before the crew could destroy it.[8] British pilots flying against the night-raiding Zeppelins learned how effective the blackout regulations were. They had at first no navigational aids other than the stars and, until their instrument panels were lighted, had trouble merely staying over the land. The fighter pilots were often unsure of their positions unless the searchlights and guns opened up against the raiders.[9]

If the fighter pilots had difficulty, the navigational task facing the Zeppelin crews was of course more formidable, for they had been in the air perhaps ten hours by the time they approached their targets. In two other branches of World War I aviation, navigation was equally crucial: long-range bombing airplanes and long-range overwater patrol aircraft. In all three, successful operations required a higher order of navigation than was essential in battlefield or home-defense flying. Using airships as long-range bombers was almost a German monopoly. The long overwater patrols were primarily a British concern. All of the industrialized belligerents, except the United States, employed bombing airplanes to attack objectives beyond the battlefield.

For the long-range operations, each nation attempted to perfect the existing navigational instruments and to develop new ones. In the vast expansion of aeronautical activity, firms with no previous experience took up the manufacture of aerial instruments. In the rush of wartime demands, most of the work was necessarily of the cut-and-try variety, empirical designs aimed at filling immediate needs. Many of the designers did not appreciate the degree of ruggedness that had to be built into instruments for open-cockpit airplanes. Two French aviators concluded, after the war, that

> an instrument is not good unless it is simple and robust. . . . the navigator . . . comes to prefer the rustic [instrument], which fears neither rain nor sand, to the precise, laboratory device, which will not function because the storage battery has failed or the filament of the electric lamp has broken. The sand and the rain penetrate like envy into every clockwork mechanism, and one has to adjust small controls with either numb fingers or bulky gloves . . . . Those who work in laboratories forget [such things] too easily.

The designers did learn that, no matter how rationally they proceeded with their business, the pilot was human and possessed perhaps more than the usual idiosyncrasies. An American designer reflected on his experiences:

> If the instrument for any reason fails to appeal to the individual pilot, he will take great chances rather than trouble to look at it. On the other hand, if the instrument pleases his fancy, he may grow so attached to it that he will claim he could not fly safely without it, even though the instrument be scientifically known to be inaccurate. Curious examples of this circumstance were found in the popularity of the earlier liquid type Pitot tube among the British pilots and the spinning-top inclinometer among the French.[10]

The improvement of navigational instruments, the development of new ones, and their use in combat are the subject of the remainder of this chapter.

Before dealing with the instruments that were installed in the aircraft, a piece of personal equipment that was to become almost the hallmark of the aerial navigator must be mentioned. The dead-reckoning computer, an analog device for solving the wind triangle, had appeared in 1910 (above, p. 40). It does not seem to have come into widespread use until the World War. The early navigational textbooks described only the graphical solution, drawing the wind triangle to scale on graph paper.[11] As long as aviators had no information on the winds aloft before takeoff, there was little point in solving the wind triangle on the ground. But by the time long-range flights were flown during the war, weather stations could provide at least an informed estimate. Then the tediousness of drawing a graphical solution before each flight, and whenever wind, airspeed, or course changed during flight, became apparent, and instruments were designed to solve the problem more quickly.

The dead-reckoning computers took two general forms. One followed the lines of the 1910 French computer, which was designed to be used on a chart. The desired course, drawn on the chart, formed one side of the wind triangle; the computer included two adjustable rulers to represent the other two sides, the airplane heading and airspeed vector, and the wind vector. Such an instrument could be used for planning a flight on the ground, but manipulating it on a chart in the cockpit was not so easy.[12] Computers of the second group contained a rotatable circle, ruled with perpendicular lines, that served as a graph. The computer was thus freed from the chart and could be used equally well on the ground or in the air. At the end of the war, the British had developed the most advanced computer of the second type in their Course and Distance Calculator.

The British computer was derived from a naval instrument used to work problems involving simultaneous velocities, such as the interception of a moving ship. The naval instrument was too large for use in the air, and the scales were not appropriate for air work. The RAF computer was 9 inches in diameter. The graph and the two movable arms were positioned to represent the known values of the wind triangle and clamped in place with a central thumb screw. The unknown parts of the triangle could then be read against the graph. A circular slide rule was added around the periphery of the computer to solve the navigational problems that can be expressed as proportions, such as those involving time, speed, distance, and fuel consumption. For example, if an airplane is flying 103 mph, how far will it go in 14 minutes? By setting 103 over 60 (minutes per hour) on the slide rule, the answer, 24 miles, appears over 14. A separate set of scales to convert the indications of the airspeed meter (indicated airspeed) to the actual velocity of the craft through the air (true airspeed) would be added to the computer after World War I; at first the airspeed scales were available only as a separate computer.[13]

After moving the wind triangle problem from the chart to the computer, courses and distances still had to be measured on the chart. In using ordinary draftsman's implements in the cockpit, with the chart mounted on a rigid board, protractors and rulers were often broken or lost overboard. The aviator frequently needed to wear gloves, which meant that his utensils had to be larger, more rugged, and if possible designed for use by only one hand. Such needs led to the production of more elaborate chart boards, equipped with protractors, parallel rules, and distance scales, all securely attached and designed for convenient storage in the cockpit.[14]

Most Probable Position

A collection of navigation computers belonging to Captain P. V. H. Weems. *Weems & Plath photo.*

Navigation in World War I

Of the instruments installed in airplane cockpits in 1914, the magnetic compass was the most important for long-range flight. Something of the instrument's prewar development has been described. Its behavior was still completely unsatisfactory during turns, and the British Advisory Committee for Aeronautics, the Compass Department of the Admiralty, and the Royal Aircraft Factory had begun a thorough investigation of the compass early in 1914 (above, p. 47). The work was under the supervision of Keith Lucas at the Factory. Theoretical investigation, laboratory experimentation, and flight testing (in the course of which Lucas was killed) throughout the war resulted in a series of improved compasses and, perhaps more important, an understanding of certain inherent limitations of the instrument.[15]

Lucas was the first to explain the odd behavior of the compass during a turn. Because the earth's magnetic field is not parallel to the surface of the earth but dips down toward the nearer pole, the north-seeking end of the magnetized needle is pulled downward in the Northern Hemisphere. Only the horizontal component of the earth's field is useful in finding direction, and to keep the needle horizontal, the downward force on the northern end of the needle is countered by placing the pivot off-center. The pull of gravity on the longer southern end balances the magnetic pull on the northern end, and the compass card remains level. But when an aircraft turns, it tilts sharply out of the horizontal and carries the compass card with it. The magnetic needle is then placed in a different attitude with respect to the earth's magnetic field. The vertical component of the earth's field comes into play, causing an error that varies with the aircraft heading and direction of turn. The greatest error occurs when turning from north to east. By altering the magnetic characteristics of the compass, Lucas was able to reduce the turning error but could not eliminate it.[16]

The compass that resulted from Lucas's experiments was the R.A.E. Mark II. It had a much longer period than previous airplane compasses, which meant that after a turn, it came to rest on the new heading slowly. Such a compass was more useful for a navigator than for a pilot, who needed to know the new heading as soon as possible after a turn. Therefore another compass, the Creagh-Osborne type 5/17, was designed at about the same time. Having a shorter period, this compass was more suitable for fighter airplanes, and it was widely used during the war. In the United States it was copied as the General Electric type B.[17]

In the last year of the war, G. R. C. Campbell of the Admiralty Compass Observatory and G. T. Bennett of Cambridge University combined the best features of the R.A.E. Mark II and the Creagh-

Osborne 5/17 to produce the model 6/18 aperiodic compass. Replacing the card with a spider of radial wires reduced the inertia of the compass. During a turn, it began to move to the new heading quickly but did not move so fast that it overshot the correct heading. The compass therefore did not oscillate; it had no period—hence the name, aperiodic. The Campbell-Bennett compass was the first of a long line of compasses used by the British and many other countries after World War I.[18]

In Germany the most significant wartime compass development took a different direction altogether. No matter how carefully airplane designers avoided the use of magnetic metals, the engine produced a significant magnetic disturbance. Bombs and machine-gun ammunition contained large amounts of iron, and these items were expended during combat, so that deviation from this source changed during a single flight. The Germans therefore devised a completely new compass that could be mounted as far as possible from engine, bombs, and ammunition. They put the sensing element in the tail of the airplane and provided a remote indicator in the cockpit.

The problem in designing a remote compass was to find some method of picking the information off the magnetic element without introducing an error. The magnetic force was too slight for any mechanical linkage to be used. The Carl Bamberg instrument firm in Berlin found a solution in the use of selenium. The electrical conductivity of selenium increases when light strikes it. The sensing element of the Bamberg compass was designed so that a shutter allowed a light to strike one of two selenium cells whenever the airplane was off the desired course. Electricity flowed through the selenium to an ammeter in front of the pilot. The ammeter needle, normally centered, was deflected to the right or left, showing the pilot that he was off course. The pilot turned in the direction indicated by the instrument until the shutter cut off the light from the selenium cell. No current then flowed, and the ammeter needle moved back to the center, on-course position. The navigator set the desired course by positioning the shutter with a flexible shaft. The instrument was particularly suitable for multi-place airplanes, because a number of repeaters could be connected to a single sensing element. At the end of the war the Bamberg compass seemed to be one of the more promising instruments for use by civil aviation.[19]

Because the turning error could not be eliminated, none of these compasses gave an accurate indication during a turn. But by explaining the cause of the error, Lucas largely restored the instrument to aviators' favor; thereafter they would know when to trust the compass and when not. An entirely different instrument was needed for con-

Most Probable Position
72

RAF course and distance calculators.
Bureau of Standards photo, courtesy of Capt. Weems.

trolling the airplane in a turn, and several turn indicators appeared during the war.

The earliest British instrument sampled the air pressure at each wing tip to detect turning motion. The gyro was pressed into service by the Germans and the French to perform a similar function. A gyro rotating at high speed, with its equator parallel to the horizon, retains that attitude even if the airplane carrying it does not. The gyro therefore served as an artificial horizon.[20] By reference to the instrument, a pilot could hold the wings level. The turn indicator and the artificial horizon were not navigational instruments as such, but the use of one or the other was essential for accurate navigation at night and in weather; only with their help could a pilot hold a steady compass heading.

The magnetic compass of 1918 was a significantly better instrument than the 1914 model had been. Progress with the airspeed indicator and pressure altimeter was less dramatic.

The most promising airspeed meter at the start of the war used the pitot tube. The instrument compared the impact air pressure with the static air pressure by applying them to opposite sides of an elastic diaphragm. At low speeds the pressure differential was slight. Early in the war, therefore, the pitot was replaced by the venturi tube, because the latter produces a larger pressure differential. A venturi resembles two funnels, with the small end of one connected to the small end of the other. Air moving through the tube must move faster in the constricted section, and a partial vacuum or suction results. The difference between the suction of the venturi and static pressure exceeds the pitot-static difference. Comparing the positive pitot pressure with the negative venturi pressure provides a still greater differential. By 1918 instrument designers were using all three combinations of pressure heads: pitot-static, venturi-static, and pitot-venturi. A second major difficulty was the choice of material for the diaphragm. Doped silk and rubber were used in early airspeed meters, but neither proved durable. Thin, corrugated metal held up better but gave a smaller motion for a given pressure differential. A series of small metal diaphragms, each exposed to the same pressures and connected by springs, proved more satisfactory than a single large one.[21]

Improvements of design details continued throughout the war. Nevertheless, the first Technical Order published by the United States Army's Division of Military Aeronautics in October 1918 said: "Air speed indicators are reported to be inaccurate. Efforts are continually being made to improve these instruments." And that is all the Technical Order said on the subject! Shortly thereafter, two French aviators wrote of an instrument that employed a venturi: "To fly at '120 of the

Six early U.S. Navy aircraft compasses.
U.S. Navy photo, courtesy of Capt. Weems.

Badin' (to use a current expression), that is to say, piloting so that the hand of the speed indicator remains constantly on 120, does not signify anything . . . . indications vary from one airplane to another, depending on the way the indicator is mounted and whether the tube is partly stopped up, deformed, the rubber joints leak . . . ." One American instrument manufacturer gave up on the problem of the diaphragm and reverted, in 1920, to the earlier liquid-filled tube, but this was a technological dead end. Generally, airspeed indicators made considerably less progress than did compasses in this period.[22]

Altimeters, on the other hand, needed less improvement. The main alteration required was to extend the instrument's range. In early 1915 the General Aeronautical Company, Ltd., offered a new line of light, "accurate" altimeters, graduated at intervals of 100 feet to a maximum of 10,000, with "red figures, from 8,000 to 10,000 ft., indicating altitudes at which one might be considered safe from shell fire." (Both pocket and wrist models were available; "they should form

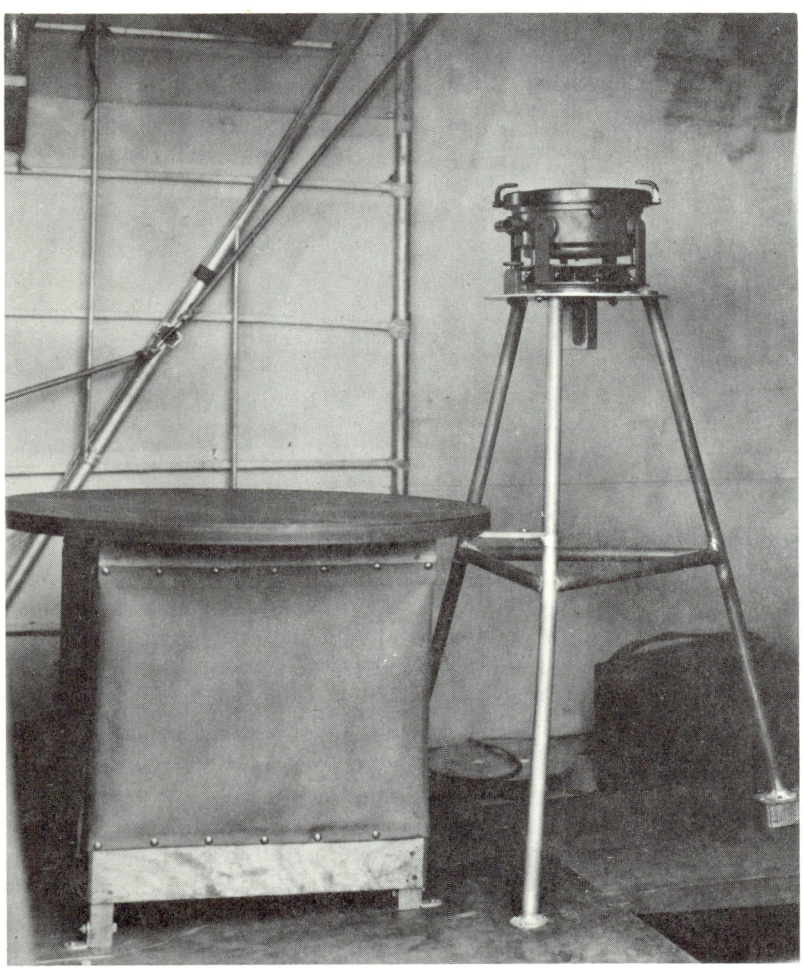

Aperiodic compass installed in Fokker Tri-Motor, June 1927. *USAF photo.*

most acceptable presents to any of our flying officers at the front.") Within two years aircraft were flying above 20,000 feet and were sometimes hit by antiaircraft fire at those altitudes. Extending the range of the altimeter required multiplying the motion of the pressure element even more than before, because the pressure change per thousand feet decreases with increasing altitude. The pressure element of a typical altimeter flexed 1 mm in a pressure change from sea level to 20,000 feet; this motion had to be multiplied 200 to 800 times, before being transmitted to the pointer.[23]

The final navigational instrument that had been used in prewar airplanes, the driftmeter, received extensive attention before and during the war, because the driftmeter supplied much of the information required for accurate bombing. Before anyone tried it, dropping bombs onto stationary surface targets seemed easy enough. It was tried as early as 1910, and the difficulties became apparent immediately. For a time the airship seemed the only choice for an aerial bombardment vehicle, because it could hover over its target, correcting the aim from bomb to bomb much as artillery did, until it achieved a direct hit. The French staged a bombing competition for airplanes in 1912, and Lieutenant Riley Scott, USN, surprised the military community by his unexpected success. From 2,600 feet, 8 of his 15 bombs hit a target of 124 x 375 feet. From 650 feet, he did better—12 of 15 bombs into a 66-foot diameter circle. By the eve of the war the bombing problem was generally understood. Designing instruments to collect the required data, combine it, and present the result in such a way that the bomb could be aimed became a continuing concern for all air forces.[24]

At the moment of release, a bomb has the velocity and direction of the airplane carrying it, that is, the airplane's groundspeed and track. In bombing from level flight, the bomb has no vertical velocity at the moment of release; but gravity immediately imparts a vertical velocity, which is opposed by the resistance that the air offers to the bomb. Air resistance is a function of the shape of the bomb case and is constant for a given type of bomb. The time of fall of the bomb therefore depends on two things: the altitude of the bomber over the target and the characteristics of the bomb case. The bombsight must determine the horizontal distance and direction from the target to release the bomb, so that in its time of fall, the bomb's horizontal motion will carry it exactly to the target. In World War I the bombing problem was simplified whenever possible by flying directly into the wind, or else with the wind, so that there would be no lateral drift to contend with.[25] Then the point at which the bomb should be dropped could be expressed as a sighting angle, measured from the vertical as zero, increasing forward in the direction of the airplane heading. The higher the groundspeed of the bomber, the greater the sighting angle for bombing. At speeds and altitudes common in World War I, the bomb might be dropped when the target was 10° to 20° forward of the vertical. The driftmeter, or an instrument similar to it, could provide an important piece of information for bombing: the initial velocity of the bomb, which was the groundspeed of the bomber. The bombsight, then as now, could compensate only for the effect of the wind at the bomber's altitude. The wind usually varies with altitude, so that a bomb is affected by different winds as it falls. There is no really satisfactory solution to this problem.

The first driftmeters used in airplanes combined a magnetic compass with an optical system that carried a grid (above, p. 51). One set of grid lines was for measuring the airplane's drift, the other set for groundspeed. Because the grid was attached to a magnetic compass, the picture of the earth's surface viewed through the grid was seldom steady. To improve the accuracy of both drift and groundspeed measurements, most driftmeters designed in World War I and afterward eliminated the compass element. Nevertheless, the motion of the airplane alone still caused the grid to move back and forth over the landscape, and in rough air the driftmeter was frequently unusable.

Early in the war a number of hybrid driftmeter-bombsights were produced. They combined the driftmeter grid (lines for drift, others for groundspeed) with a sighting telescope and cross hairs, to which was attached an index for measuring the angle made by the telescope with the vertical. With most of the sights, the airplane was first turned directly up- or down-wind. The groundspeed was then measured, the sighting angle found from bombing tables, and the telescope was set at that angle. When the target appeared under the cross hairs, the aviator dropped "the bomb over the side of the *nacelle* or *fuselage*."[26] Later, release mechanisms, controlled from the bombing station, came into use. Some sights were calibrated so that the telescope could be positioned directly from the groundspeed, or even from the time interval taken by an object crossing from one of the timing lines to the other. In either case, the sighting telescope was positioned at the correct angle, while time was saved by eliminating one or two computations.

The sight could seldom be located in the pilot's compartment because during the last critical moments of the bomb run, the pilot could not see the target after it passed beneath the nose of the airplane. In some early bombers, the sight was placed in the floor of the fuselage, near the bomb bay. The preferred position came to be the nose of the airplane, the position that offered the best visibility. The bombardier—who might be another pilot, navigator, or gunner; a specialist to do only the bombing was not part of the World War I crew—had to be able to communicate with the pilot during the bomb run, to tell him which way to turn and how far. The most primitive methods, depending on the location of the sight relative to the pilot, were hand signals or slapping the pilot on the legs. Later, signal lights, controlled by the bombardier, were mounted on the pilot's instrument panel. The bombardier signaled for a left or right turn by pushing a button and holding it until he wanted the pilot to level the wings. The Germans used more finesse. One of their early bombsights incorporated automatic directional signals to the pilot. If the bombardier, in placing the cross hairs on the target, had to push the telescope to the right or left, an electrical contact closed, lighting one of two small lamps in front of the pilot. One of the "giant" German bombers, under design as the war ended, included a duplicate set of flight controls at the bombing station, so that the bombardier could make slight course corrections himself.[27]

In all bombsights and driftmeters, the optics needed to be held precisely horizontal, first when computing drift and groundspeed, later when sighting on the target. If the bombsight optics in an airplane flying at 10,000 feet happened to be 1° from the horizontal at the moment of bomb release, the bomb would strike the ground 175 feet from the target—with World War I bombs, "a complete miss." Because no airplane would fly perfectly level, the optics of the sight needed to be stabilized with reference to the horizon. The problem was identical to that of the artificial horizon for the sextant or the turn indicator for the pilot. German bombsights incorporated an air bubble, and the bubble was subject to the same errors as the bubble in the sextant: it responded to all aircraft accelerations, as well as to gravity. In the United States, Elmer Sperry attached his gyro to a bombsight in 1915, but the instrument does not seem to have been successful.[28] Gyro-stabilized bombsights would reappear between the wars, after the gyro had been further improved.

The altitude above the ground had to be known for accurate bombing, both to determine the time of fall and to compute the groundspeed of the bomber. There was no better means for computing altitude than to subtract terrain elevation, taken off the chart, from the height indicated by the aneroid barometer. This procedure had to be used throughout the war to find altitude for bombing, but a method was found for computing groundspeed without knowledge of altitude, by first determining the wind with the driftmeter.

If a navigator knows only the airspeed and heading of his airplane and the drift he is experiencing, he cannot compute the wind. Consider the airplane headed south from point A in Figure 10. A northwesterly wind would cause his airplane to drift left, to the east. If, however, a second airplane left point B, headed east, the same wind would produce a right, or southerly, drift. Neither navigator could compute the precise wind, because neither would have enough information to solve the wind triangle. But if they pooled their data, they

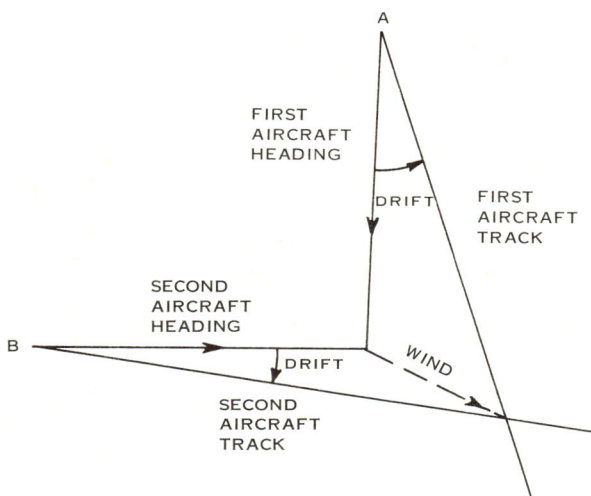

FIG. 10. Computing the wind from drifts on successive headings. Two aircraft flying different headings experience different drifts from the common wind. By drawing the two wind triangles together, with the wind vector as a common side, the triangles may be solved graphically. The heading, airspeed, and drift of both airplanes are known. Heading plus drift gives ground track. The intersection of the two heading lines connected to the intersection of the track lines equals the wind.

could solve both triangles, because one side, the wind vector, is common. There is only one possible wind that will produce two precise drifts on two different headings, and it can be found graphically as illustrated in Figure 10. The principle also applies if only one airplane flies two or more headings in succession. After measuring the drift on each heading, the navigator combines them to find the wind. Knowing the wind, with the airspeed and heading, he can then find groundspeed—with no knowledge of the airplane's altitude.[29]

The first driftmeter designed for this method of wind computation was the invention of Lieutenant Colonel Crocco of the Italian Air Force. A Royal Air Force officer, Major Harry Egerton Wimperis, soon produced a better instrument on the same lines; he called it a wind-gauge bearing plate. Both combined a driftmeter with a dead-reckoning computer. The instruments first measured the drift on each of a sequence of headings and then combined the drifts, headings, and airspeed on an analog computer to find the wind, a procedure which the British called a wind star, from the pattern of crossing lines formed by plotting the sequence of headings and drifts. Americans now call it a double-drift wind. The method was especially useful for aircraft flying patrol patterns that incorporated a number of turns. The navigator could measure the drift before and after each change of heading and compute the wind.[30] For other aircraft, it might be necessary to turn off the desired course for perhaps a minute, but this was a small price to pay for an accurate knowledge of the wind, and therefore of the groundspeed.

The Crocco and Wimperis driftmeters were elaborate instruments, too bulky for use in any but the largest airplanes. Smaller, simpler driftmeters were still required. For the single-seat airplane, a set of wires, radiating at 10° intervals from a point on top of the instrument panel, served the purpose. The pilot watched the movement of objects on the ground with reference to the wires and, with practice, could estimate drift to about 2°. For the two-seat airplane, Elmer Sperry designed a driftmeter that was connected to the pilot's compass. As the observer rotated the driftmeter to align the apparent motion of the earth with the drift lines, cables displaced the lubber line of the pilot's compass right or left an equivalent amount. The pilot could continue to steer the same heading, in degrees, but would actually be steering a different direction in response to the observer's positioning of the driftmeter. Sperry pointed out the convenience of the arrangement: "This is found very practical, inasmuch as it vastly simplifies the pilot's operation . . . . The fact that the lubber line is displaced . . . is something with which the pilot has nothing to do and is not concerned—he simply continues on his original compass course." Unfortunately, a pilot should know the direction he is flying; and tampering with the compass, making it indicate the wrong direction, was a dangerous expedient. The instrument does not seem to have had a long life, and the principle was not copied in later models. A more successful American driftmeter, produced by the Pioneer Instrument Company, was modeled on the Crocco instrument, with the analog computer omitted. The navigator used the Pioneer driftmeter only to measure drift and then solved the wind triangle on a small, hand-held, dead-reckoning computer.[31]

Because the driftmeter was so closely associated with the bombing problem, the instrument received much attention from the designers. The progress made with it during the war was comparable to that with the magnetic compass. Only one entirely new navigational instrument appeared during World War I: directional radio.

Radio and powered flight belong to the same technological generation; they have grown up together. Marconi transmitted messages across the Channel in 1899 and across the North Atlantic in 1901. By 1908 messages had been received and transmitted by balloons in

Airplane equipped with wireless, 2 November 1912, Fort Riley, Kansas. The aviators are Lieutenants Follett Bradley and Henry H. Arnold.
USAF photo.

France, Belgium, and Germany. By 1910 or 1911 messages were being transmitted from airplanes in France, Germany, Great Britain, and the United States.[32] Ever since, radio has been closely associated with aviation.

A means of communication between the aerial vehicle and the ground was certainly needed, but as early as 1908 the possibility of using radio as a navigational aid was being explored. A French balloonist, thinking as always of being blown out to sea, suggested that radio stations be erected along the coast. If one heard signals from such stations, which would have a range of about 25 miles, he would be warned of the danger. "The radio beacons along the coast would warn balloonists of the sea, just as light beacons warn the sailor of the land." A German theorist went further, suggesting that about 100 radio stations be evenly spaced throughout Germany. A balloon would be in range of one or more all the time. By identifying the station, or stations, that he heard, a balloonist would know his approximate position.[33] Neither scheme came to anything, because both presupposed that the range of a radio station was constant and that

Two-way radio-telegraph installed on Glenn L. Martin airplane #50, fall 1916, San Diego. Note the antenna reel.
USAF photo.

the stronger of two signals necessarily came from the closer of two stations. Experience with radio dissipated those false assumptions. But for communication, the value of radio became more apparent as World War I approached.

If aircraft were to be of use in reconnaissance or artillery spotting, rapid communication with ground forces and headquarters was essential; and radio was obviously preferable to dropping notes or flashing lights. Efforts to adapt radio to aircraft accelerated as the war approached. The first problem to overcome was antenna size. Before the advent of the amplifier tube, receivers were extremely inefficient. The antenna had to deliver a strong signal to the receiver if the latter was to produce an audible signal, and this could only be done by building a large antenna. Two antennas are required in radio. Surface stations use the earth (the "ground") for one and could employ large structures for the other (the "aerial"). Both antennas had to be "aerial" on aircraft. A common solution was to string antennas between wing tips, nose, and tail. A longer antenna could be provided by the use of a trailing wire, rolled up on a windlass for takeoff and landing; but as long as pusher propellers were in general use, pilots looked with disfavor on trailing wire antennas. By 1914 the metal parts of the airframe were sometimes connected to serve as one of the aerials, but before the era of all-metal airplanes, this method gave poor results. As in so many other ways, the airship had an advantage over the airplane. Antennas of any desired length could be wrapped around the gas bag or trailed behind the aft propeller with no danger of fouling.[34]

The inability to provide antennas as large as those in use on the ground, the noisy environment of the airplane, and the electrical interference ("static") generated by the ignition of the engine, threatened for years to restrict airplane radio to one-way traffic. The airship could stop its engines and drift while receiving a message, but the problem could not be similarly avoided in airplanes. Before the war, reception was limited to ranges of only a few miles and was not dependable even then. The solution was found by building the speaker into a helmet that shut out much of the noise, electrically screening the receiver from magneto interference, while increasing the receiver output many times by the use of amplifier tubes.[35]

Size and weight of the early radio equipment also hampered its use in airplanes. Successive models of radios tended to be both more powerful and less heavy than their predecessors. Nevertheless, at the start of the war radios weighed about 70 pounds, a significant burden for the airplanes of the day. Most two-seaters could not lift a two-man crew, radio, and useful load of fuel. Further, the rear cockpit was often needed to store the radio equipment until designers learned to provide space for the black boxes.[36]

While radio was being developed for communication and being adapted to use in the air, its directional qualities were investigated. Radio waves radiate in all directions from a simple wire antenna. While radio engineers were having so much trouble getting sufficient power to the receiver, such a radiation pattern seemed wasteful. The energy would be better used if concentrated in the direction of the receiving antenna. As the number of radio stations increased, mutual interference also became a problem. For both reasons, radio engineers began to experiment with directional antennas, those which receive and transmit better in certain directions than in others. From this point it was but a step to the use of directional radio as a navigational aid for ships, and later for aircraft. By 1914 two distinct methods of position finding by radio were in marine use. One, the Bellini-Tosi, employed a directional receiver; the other, the Telefunken compass, a directional transmitter. Both instruments made use of the same principle: an antenna in the shape of a loop, triangle, or square will transmit or receive the greatest energy in the plane of the antenna. It will transmit or receive the least energy in the direction perpendicular to the antenna.[37]

The Bellini-Tosi instrument used two large, fixed, triangular antennas, mounted at right angles to each other and perpendicular to the ground. A small, rotatable coil, attached to an azimuth pointer, was placed in the center of the large antennas. When an incoming radio signal struck the large antennas, the antenna that was closer to being parallel with the direction of the transmitter picked up the stronger signal. The two antennas reradiated the received signal in a spatial pattern that corresponded to the direction of the transmitter. The reradiated signal was picked up by the small, rotatable coil in the center. By turning the small coil through 360° and noting either the strongest or the weakest signal, the operator determined the direction of the incoming radio waves. The weakest signal would be received when the coil was perpendicular to the direction of the transmitter, the strongest when the coil was parallel to the direction of the transmitter. The minimum signal position was more distinct than the maximum; finding the position of the minimum signal, or the null, was therefore the preferred method. The equipment could not differentiate between signals coming from exactly opposite directions; transmitters east and west of the directional receiver caused similar indications. Because of this 180° ambiguity, the rotatable coil was geared to a double-ended pointer. The operator, after adjusting the coil for minimum signal, read either the direction to the transmitter or the opposite direction. The Bellini-Tosi equipment could be installed either ashore or on board ships. In the latter case, the marine navigator could get a bearing, or line of position, from any transmitter. The 180° ambiguity caused no confusion if the ship was approaching a coast, because one of the two possible directions would usually place the ship inland and could therefore be discounted. Working with island stations or transmitters on other ships, the ambiguity could be a source of difficulty. The Bellini-Tosi equipment originated in Italy and France. Britain's Marconi later contributed to its improvement, and the Marconi Company marketed the device in Britain and the United States before the war.[38] In Germany, the Telefunken Company attacked the problem from the other end and devised a directional transmitter.

The Telefunken compass consisted of a number of concentric directional antennas. In the first model there were 16 antennas, producing 32 directional signals, one for each point of the compass. During the war the number of antennas was increased to 90 for greater accuracy. In operation, a nondirectional signal was transmitted, followed at precise intervals by a brief transmission on each directional antenna in turn. Any ordinary, nondirectional receiver could be used to get a bearing from the Telefunken compass, but the navigator had to have a watch that was synchronized with the transmitter. On receipt of the nondirectional signal, the navigator started the special watch. On hearing the minimum signal, he stopped the watch but, instead of time, read the bearing to the station (or the reciprocal; ambiguity existed in this system also).[39]

The two systems were being installed on the German, French, and British coasts at the outbreak of the war. Each had strengths and weaknesses. A navigator always prefers to control the source of information so that reliance on outside agencies is minimal. The Bellini-Tosi, if installed on the ship, allowed such independence, for the navigator measured the bearings himself, or had the ship's radio operator do so. Any transmitter, if its position was known, could be used for a bearing. The ironwork on the ship reradiated received signals, just as the main antennas did, and thus introduced an error into the rotatable coil used for measuring direction. This error could be measured by careful calibration, and thereafter readings could be corrected, much as the navigator applied deviation corrections to magnetic compass indications.[40] If the Bellini-Tosi equipment was located on shore, a marine navigator had to request a bearing from the shore station. That meant relying on an unknown person, and it also required both a transmitter and receiver on the ship. With the Telefunken compass, the navigator did his own work. He needed only a receiver but was restricted to using only the special Telefunken stations. Further, he had to have a stopwatch synchronized with the Telefunken equipment. With the passage of time, the Bellini-Tosi became more accurate through successive improvements. Little could be done to improve the Telefunken accuracy, except to add more antennas, which reduced the angle between successive signals. When the Germans needed long-range direction finding for their Zeppelin operations during the war, they built ground stations similar to the Bellini-Tosi instrument.

On the eve of the war, many of the problems associated with the use of radio in aircraft had been solved. The sets were still too bulky for smaller airplanes, but miniaturization proceeded rapidly. The 70-pound radio was common in 1914; the British introduced a set weighing less than 20 pounds in the fall of 1915.[41] The theory of radio direction finding was well understood in 1914, and ground stations were being built. Some ships carried direction finders, but aircraft could not yet do so. However, having communication radio on board, aircraft could ask for bearings from the ground stations. This was the first application of direction finding to aerial navigation.

A single ground station provided only a single bearing, and two or more were necessary for a definite position. An aircraft could call a series of ground stations in turn, but a more efficient way was to organize the ground stations into groups or nets, connected by telegraph or telephone lines. One station, the net control station, answered all radio calls from aircraft. All stations in the net took bearings on the aircraft simultaneously and reported them to the net control station, where they were plotted. Two bearings provided a fix, but it was safer to use three. If they crossed at a common point or nearly so, the position was accurate. If the three bearings did not intersect at a common point, one at least was in error. By plotting the bearings at the net control station, they could be evaluated before reporting the position to the aircraft. Some navigators preferred to plot the bearings for themselves, and bearings rather than positions were sometimes reported to the aircraft. During the course of the war, the British and French erected direction-finding nets along the Western Front and along the North Sea, English Channel, and Irish Sea coasts. The Germans had a series of stations on their North Sea coast, stretching from the Danish border to occupied Belgium. The United States Navy installed a net on the East Coast, covering the approaches to New York City.[42]

Two serious disadvantages resulted from the use of ground direction-finding stations. First, an aircraft had to transmit to use the service. This automatically informed the enemy of the aircraft's position. A series of requests for position established the course of the aircraft. The number of different aircraft requesting positions and their call signs revealed one's plan to any patient, imaginative listener. The British developed this branch of intelligence to a fine art during the war and used it to arrange their defenses. A second disadvantage was the limited number of aircraft that could use the net. The quality of radio transmission was poor with early equipment. It was standard procedure during World War I to transmit every message two or three times. The aircraft also had to transmit a continuous signal for perhaps half a minute to give the ground stations a constant signal to measure. "With eight or ten ships in the air at the same time, all asking for bearings every hour or oftener, the pandemonium in the ether—the 'Battle of the Wireless,' as the radio operators called it—sometimes beggared description."[43] Both disadvantages could be overcome by having the aircrew take the bearings. In the winter of 1917-18 the Germans reverted to a directional transmitter on the ground, which the Zeppelin crews could use with nondirectional receivers.[44] The British turned their research toward placing the directional antenna on the aircraft and had succeeded before the end of the war.

The first solution found for mounting a directional antenna on a biplane was to wrap the antenna laterally around the wings. A single wire was run from the fuselage, along the top plane, down the outboard interplane strut, along the bottom plane, up the strut, and along the top plane back to the fuselage, forming a closed loop. The antenna could not be rotated in taking a bearing—but the entire airplane could. This arrangement was tolerable for homing on a radio station, that is, flying directly toward the station. For use with stations not along the airplane's course, the airplane had to be turned off the desired heading. The pilot had to make many changes of heading, until the radio operator found the minimum signal, or null, position. "The method . . . had the objection that one could not set a course and keep it. The machine had to be swung about and the navigating people objected to this," because keeping account of the various headings was difficult and the dead reckoning deteriorated.[45] The wing antenna was most useful returning to base after a mission, when the navigational problem was one of "homing."

At the end of the war the British were developing a rotatable coil antenna, about 3 feet in diameter, that could be used inside the fuselage. Such a small, relatively inefficient antenna was usable only with the amplifier tube, which increased the efficiency of the receiver many times over that of prewar equipment. The British also found it necessary to abandon the measurement of the null in airborne direction finding. Ground stations measured the minimum signal because its limits were more distinct than the maximum. In the air, the high noise level often blotted out the null, and the procedure was therefore reversed. The radio operator sought the maximum signal and bisected it, to find the bearing. Finally Captain James Robinson of the RAF devised a method that combined the advantages of both maximum and minimum signal methods. He used two coils, mounted perpendicularly to each other and rotatable as a unit, and he provided a switching arrangement so that the operator could attach one or both of the coils to the receiver. One coil was used to find the approximate maximum signal. The second coil, which would thereby be positioned at the approximate null position, would then be added to the circuit along with the first coil. If the second coil was in the precise null position, its addition to the antenna circuit would not change the signal, and the operator read the bearing. If the signal level changed with the second coil in the circuit, the coils had to be moved slightly until the exact null was found by trial and error. Robinson thus combined the advantages of the maximum signal method, which produced a signal loud enough to be heard above the airplane noise, with the more precisely defined limits of the null position, which produced a more accurate bearing.[46]

The range of direction-finding equipment increased during the war from less than 25 miles to a few hundred, due primarily to the introduction of the amplifier. The longer ranges brought unexpected difficulties with them. Before the war direction-finding bearings were considered to be exact, and with the short ranges then possible, no error was apparent. With increased ranges, gross errors appeared. The British were in an excellent position to recognize the magnitude of these errors, because they regularly intercepted messages between the raiding Zeppelins and the German ground stations. At the same time the British had an elaborate network of ground observers who reported the position of the Zeppelins. The British therefore were able to evaluate the accuracy of the German direction-finding equipment better than the Germans could themselves—and the British determined that they had to have a higher order of accuracy for their planned airplane raids on Berlin. With research centered at RAF Station Biggin Hill, south of London, and at Orfordness, on the Suffolk coast, the British made a start on the extensive experimentation necessary to isolate and correct the various errors to which direction finders are prone. They thought they had devised techniques to allow radio navigation with an accuracy of 5 to 7 miles.[47] The war ended before the raids could be flown, and the equipment was not tested in combat. Considering the vast amount of work done on radio direction finding between the wars, it seems likely that the Berlin raids would have provided the Germans with the spectacle of airplanes groping for their targets, much as the British watched the Zeppelins throughout the war.

The Zeppelin raids against England, one of the more dramatic aspects of World War I aviation, posed severe navigational problems, which were never really solved. The Zeppelins first operated out of advanced bases in occupied Belgium, but Allied airplanes could reach those bases too easily. In 1915 the big ships were withdrawn to bases on Germany's North Sea coast. From there, the shortest route to their targets was across the North Sea, passing to the north of neutral Holland. The German crews at first simplified their navigation by staying in sight of the Dutch islands as long as possible, but some airship commanders came to believe that spies in Holland were warning the British of inbound raiders. To preclude such early warning, these commanders either flew straight out to sea from the German coast, or else took the longer route south of Holland, crossing occupied Belgium.[48] Whatever the route, the airships had to travel about 350 nm to reach London or the industrial targets of the English Midlands.

The prevailing winds over the North Sea are westerly—head winds for Zeppelins approaching England. The airships cruised at airspeeds of 45 to 60 knots. Depending on the strength of the opposing wind, time from takeoff to target therefore varied from eight to 12 hours. The Zeppelin crews knew the strength of the winds in the German Bight from a weather station at Heligoland. Another station on the Belgian coast and occasional reports from submarines provided the only information they had from farther west. Weather systems move off the Atlantic across Britain toward the continent, so that the Zeppelin crews were almost totally ignorant of the conditions to be encountered over their targets; and for more than half of each mission, the forecast winds were of no practical value.[49]

The first Zeppelin raid on England was on the night of 19 January 1915. Two Zeppelins from Hamburg flew to attack the Humber ports. Flying separately, they followed the offshore route along the Dutch islands. Map-reading fixes provided track and groundspeed as far as the island of Terschelling, from which point the airships crossed the North Sea on compass headings. Observation of the waves indicated to the commander of one of the airships, *L 3*, that the wind was more northerly than anticipated. He therefore determined to attack a target farther south. Upon crossing the English coast, which he recognized as northern Norfolk, he dropped parachute flares, pinpointed his position over a lighthouse, and followed the coast south to Yarmouth, which he bombed. The commander of the second Zeppelin, *L 4*, crossed the coast later at almost the same place but believed he was fully 80 miles farther north. Having failed to identify his landfall position, his subsequent navigation was also erroneous. He bombed various villages that were showing lights and discovered where he had been only upon reading the English newspapers several days later.[50]

The experiences of the two commanders on the very first Zeppelin raid foreshadowed with surprising accuracy the entire campaign.

Some of the commanders became adept at determining the wind over the sea, either from watching the waves or by tracking a floating flare with the bombsight-driftmeter. They were therefore able to compute reasonably accurate DR positions and, upon crossing the coast, could usually locate themselves precisely by map reading, unless clouds intervened. They could seek out their assigned targets and at least bomb the specified towns. Other commanders never seemed to get the knack of navigation, and they could only attack targets of opportunity. Considering the total Zeppelin campaign, probably half of the places that the German crews reported they had bombed were misidentified.[51] The need for a new navigational aid was apparent, and the Germans hoped that radio direction finding would fill the need.

The first two ground stations, on the westernmost German island of Borkum and at Nordholz, were placed in operation in mid-April 1915. Two radio stations provide two bearings. If the bearings cross, they define a fix; but these stations, as viewed from the target areas, were almost in line (Figure 11). For the most accurate fix, two bearings need to cross at 90°. The Germans later erected additional radio stations at Bruges and on the island of Sylt. These stations gave excellent coverage over much of the North Sea. Airships in the target areas still had to use bearings crossing at acute angles; but the Germans, having built stations on the Belgian coast and near the Danish

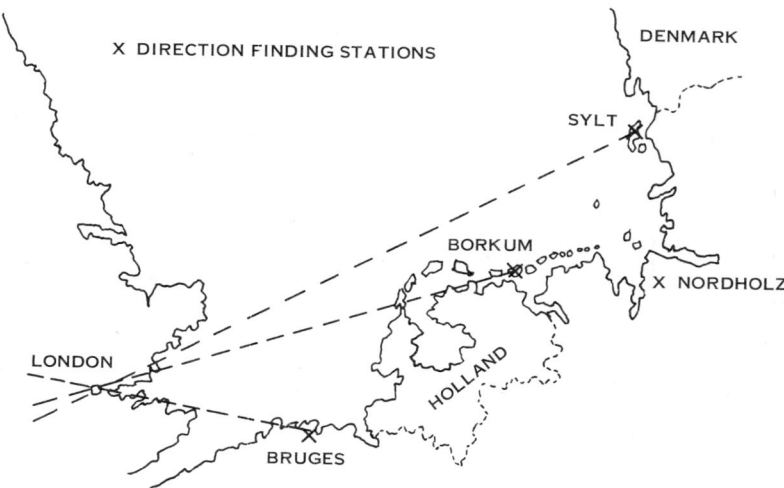

Fig. 11. The direction-finding stations used by the Zeppelins raiding England. The lines show the angles made by the radio bearings over London. For best results, two bearings should cross at 90°.

border, had run out of real estate. Their radio navigation suffered not only from less than perfect positioning of the ground stations, however; a whole series of errors affected the bearings, and these errors were not even suspected for much of the war.[52]

After the first few raids the British instituted censorship. The historian of the German naval airships assumed that because the Germans never knew just how bad their navigation was, they saw no need to change their methods. While the assumption may have been generally valid, from time to time a Zeppelin crew must have been able to identify a landmark absolutely without doubt—London, for instance, or a peculiarity of the coastline. If radio bearings placed the airship elsewhere at that time, the crew would have known that the radio bearings were faulty. Nevertheless, radio became the prime German navigational aid, because night map reading became more difficult as the war progressed. The blackout regulations spread quickly and were effective. As British defenses improved, the Zeppelins had to forego raiding near the full moon. They were also forced to ever higher altitudes. All three changes made identification of surface landmarks more difficult and offset the experience gained by the German crews in repeated raids over the same area.[53]

At times faulty navigation led directly to the loss of airships. In May 1916 eight Zeppelins attempted a raid on northern England. Six of the commanders, recognizing that the wind had increased, diverted to attack targets in the Midlands. The two who continued were over a solid layer of clouds and radio reception was marginal. One turned back in time, but the other, the commander of *L 20*, continued too long. In the middle of the night the clouds cleared and he found that he was over Loch Ness, well beyond his craft's radius of action with the existing southerly wind. He tried to reach the northernmost German base, but found he could not even intercept the German fleet in the Skaggerak. The airship was finally wrecked in Norway.[54]

The Zeppelins were not heated. As the British defense pushed the airships ever higher—above 20,000 feet before the end of the war— the liquid in the compasses sometimes froze, even though a generous mixture of alcohol was used. Oxygen for breathing was carried, but the apparatus was primitive and often malfunctioned. When the compass froze, if the helmsman was not numbed by the cold or befuddled by lack of oxygen, and if the skies above were clear, the airship could be steered by the stars. But in June of 1917 a British airplane found *L 48* wandering toward the north, long after dropping its bombs, and shot down the airship. Three members of the crew survived the fiery plunge—an event unique in the history of the Zeppelins —and blamed a frozen compass: otherwise the airship would have been well on its way toward home before the British airplane got into position.[55]

The most spectacular disaster of the Zeppelin campaign resulted directly from the German inability to forecast weather over the British Isles and the inability of the flight crews to detect wind shifts promptly in flight. On the night of 19 October 1917, eleven Zeppelins attempted attacks on the Midlands. To take advantage of less unfavorable winds, it was customary for the Zeppelins to cross the North Sea low, from 2,000 to 4,000 feet, climbing before they reached the area where they might be intercepted. Climbing always took the airships into different winds, which complicated their navigation, but on this particular night an unusual pressure system existed. Gale winds were blowing above 10,000 feet over Britain, while the winds near the surface of the North Sea had been almost calm. Only one airship commander recognized the dangerous situation and took the only safe course: he descended to escape the high winds, dropped his bombs without effect, and returned via the North Sea route. The others all fought the head winds too long, at times being pushed backward, and consantly drifting south. Of these ships, five managed to return to their bases, either by violating Dutch airspace or making the hazardous crossing of the Western Front. One of the five commanders was so uncertain where he had been that he refused to turn in a track chart with his report. Another airship also returned to Germany but, upon descending through the clouds, the ship was 200 miles to the southeast of its home base and was wrecked in the subsequent crash landing. The remaining four airships were also destroyed. One was shot down crossing the lines in France. Three were crash-landed in southern France; one of them drifted off, after impact, and carried four crewmen into the Mediterranean. The weather forecasters had obviously failed on this mission, but the flight navigation was equally inadequate. As the commanders realized that something was amiss, they swamped the ground stations with requests for radio bearings, with the result that many had long waits before receiving bearings. Most of the commanders did not understand their danger, or react to it, quickly enough; and five of the 11 ships were lost. This German disaster, followed three months later by a fire that destroyed five more ships on the ground, virtually ended the Zeppelin threat to England.[56]

If radio served the Zeppelin crews so badly, one must wonder why they did not use celestial navigation. The German balloonists' early interest in celestial navigation has been described (above, pp. 30–36). German military aviators had helped test artificial-horizon sextants. German aviation periodicals had thrashed out the various methods of sight reduction. Why then did they not take advantage of this background during the war? The answer seems to be that celestial navigation was used occasionally, but that its results were not generally superior to radio. Navigation by celestial observations was much more difficult than by radio; if a navigator had to chose between two inaccurate methods, it is not surprising that he chose the easier one.

Before the radio net was completed, the Zeppelin commanders found their positions by reference to the stars at least occasionally. Treusch von Buttlar-Brandenfels, who was the only Zeppelin commander to fly throughout the war, said of a "typical" raid during the middle of the war: "At all events, at that time we no longer required the help of the stars on our journeys. We had a much better means of guidance in wireless telegraphy." He did not mention celestial navigation anywhere else in his book, and he obviously had little use for it. The marine sextant was used sometimes for star observations at dusk, which would ordinarily occur when the Zeppelin was still flying low over the North Sea. At least one commander, Ludwig Bockholt, who made the nonstop flight from Bulgaria to Khartoum and return, used an artificial horizon sextant. His navigation of the Sahara was accurate, as he followed a route from oasis to oasis, even though the air was extremely turbulent during the day. An author commenting in 1923 on Bockholt's achievement stressed that long experience was required before satisfactory results could be achieved with that particular sextant. The amount of time devoted to training aircrew members was limited; sometimes fuel was unavailable for trials of new airships. It would appear, therefore, that celestial navigation could have provided more accurate results for the Zeppelin crews than did radio, but that the Germans either refused or were unable to pay the price required to train the navigators. Late in the war, when the Germans were using large airplanes to attack Great Britain, attempts were again made to navigate by celestial observations, but navigators trained in the technique were not available in quantity, and the airplanes too relied on radio.[57]

When the overloading of the radio direction-finding system contributed to the airship disaster of 19 October 1917, the Zeppelins did not adopt celestial navigation. Rather, the ground stations were redesigned, so that aircrews could measure their own bearings without making any transmissions. An unlimited number of aircraft could thereafter use the stations simultaneously. Accuracy was unimproved, however. The last Zeppelin raid against England was on the night of 5 August 1918. Five airships, including *L 70*, the newest in the force, attempted to attack targets in the Midlands. *L 70* was shot down by an airplane. The other crews, surely nervous after witnessing the pyre, and flying over clouds, were misled by radio bearings into

Zeppelin Staaken giant bomber. Gunner and mechanic's stations (L) and radio operator's station (R).
USAF photo.

believing they were farther west than they were. All four dropped their loads into the North Sea, as much as 65 miles off the coast.[58]

First and last, then, the Zeppelin campaign against England was marred by poor navigation. Because the most valuable result of the raids was never the physical destruction they achieved, but the withdrawal of British military forces from more productive tasks elsewhere, the faulty navigation did not detract as much from the effectiveness of the raids as one might think. The many bombs dropped in open fields and at sea could certainly have been better used, but bombs on any inhabited place served to force the British to improve their air defense force. At times, however, navigational deficiencies led directly to the destruction of airships and crews, neither of which could easily be replaced.

The Zeppelins caught the popular imagination, but airplanes dropped a much higher tonnage of bombs and were employed more widely for the task. There is no precise demarcation between battlefield bombing and attacks directed at targets farther away, but as the distance increases, the navigation generally becomes more difficult. All of the major European belligerents employed the bombing air-

Handley Page V/1500, the intended "Berlin bomber."
USAF photo.

plane; and in the best military tradition, targets were often chosen at the maximum radius of action of the airplanes, so that there was little margin for navigational error. In November 1914, by dint of detailed preparation, three Royal Naval Service airplanes bombed the Zeppelin works at Friedrichshafen, flying in daylight from Belfort, 110 nm away. (A fourth airplane was unable to get off the ground with the 80-pound bombload.) In 1917 the Gotha G. IV, the mainstay of the German long-range bomber force, attacked targets regularly at distances of about 150 nm from their bases. The British counterpart, the Handley Page 0/400, coming into service the same year, had a slightly longer radius of action. That of the trimotor Caproni, used by the French as well as the Italians, was about 200 nm. In the last 18 months of the war the Germans used a variety of "giant" bombers (R-planes) with a useful radius of 250 nm, and in 1918 the British were building the Handley Page V/1500, with a bombing radius of almost 400 nm, sufficient to strike Berlin from English bases. Of all the bombing airplanes, only the Handley Page V/1500, of which only two existed at the Armistice, rivaled the range capability of the Zeppelin.*

Daytime missions were flown throughout the war, but air defenses continually improved. On the Continent the longer penetrations

---

* Statements of aircraft performance can be misleading. Speed, load, altitude, and range capabilities are frequently established on separate flights, seldom while carrying combat loads. For bombers, range or endurance is not so important as radius of action; and within wide limits the bombload can be reduced, guns and gunners removed, and the weight so saved replaced with fuel. The radius figures I have given are generalizations from the bombing missions actually flown, excepting the H. P. V/1500, which did not see combat.

were increasingly flown at night after 1915. The English were slower in putting together their home defenses: the German airplanes had to change to night raids only in September 1917. For day flights, map reading was the primary method of navigation. Setting the bombsights required measuring the variables of the bombing equation: the heading and airspeed of the airplane, the direction and velocity of the wind, and the height over the target. Because the bombers frequently changed altitude and heading near the target when the crews were dodging defenses and identifying their aiming points, the variables had to be measured quickly once the airplane turned onto its short bomb run. Early in the war the British tried setting the sights before takeoff, based on data obtained by flying a bomber across a camera obscura at the home base. The camera obscura was an optical instrument used in training bombardiers. Early models photographed the bomber as it crossed the bombing range; later versions focused a picture of the bomber onto a chart. The position, direction, speed, and altitude of the airplane could be measured precisely, and the accuracy of an imaginary bomb drop computed. The data from a run over the camera obscura did indeed provide an accurate sight setting for the conditions existing over home base at the time of the computation; but for that setting to be valid over a target perhaps two hours later, the atmospheric pressure and the winds over the target had to be identical with those over home base. The bomber crew also had to approach the target on the same heading, and a number of reasons might make a different heading necessary. The British found that presetting the bombsight improved their bombing, but that testifies primarily to the gross inaccuracy of their earlier efforts. As formation flying became the general practice for daytime operations, numbers of airplanes had to arrive at designated rendezvous points at precise times; this further complicated the navigational task.[59]

Night raids obviously presented a more difficult navigational problem. In keeping with daytime practice, map reading was used whenever possible. Some of the crews were surprised to find how well they could see the surface at night, if there was a moon or at least clear weather. Bodies of water and outlines of forests were the best natural landmarks. Railroad tracks, if in regular use, were polished reflectors under moonlight. The countryside was far from blacked-out. Artillery barrages marked the active fronts, and the larger targets were defended by groups of searchlights, which often guided the attacking bombers. Landing fields had to be illuminated at least briefly while aircraft were taking off and landing. The brightly lighted Dutch towns were useful to aircrews flying over Belgium, and after the French and Germans agreed to treat Metz as an open city, it remained brightly lighted. A British night squadron based near Rheims used the cathedral as a night landmark: it burned "for month after month."[60]

Of more importance than such incidental sources of light, ground beacons were established to guide night fliers. The French and Italians used a network of lighthouses, each flashing an identifying code. The codes were changed periodically, so that Germans and Austrians used them for navigation with less than perfect assurance. The Germans employed a variety of pyrotechnics, as well as lighthouses. At critical positions along their bombers' route of flight, the Germans fired colored rockets or star shells. The Allied crews especially admired the "Hun green-ball batteries" that fired brilliant green shells up to 6,000 feet, to mark German airfields. The Allied crews kept careful records of the location of such batteries, for their own uses. And occasionally, when a battery was moved, an Allied crew was misled. In the last year of the war the Swiss, to help prevent unintentional violations of their airspace, set up a powerful beacon near the Swiss-German-French frontier. All things considered, the landscape was not dark. "The whole dim country was . . . twinkling with little messages, with lights and flares and friendly rockets."[61]

Although some experienced crews relied on map reading exclusively, others learned to use the compass and to navigate by dead reckoning. The wartime efforts to improve the compass have been described. Whichever compass was installed, the crews had to swing their aircraft, eliminating as much compass error as possible by compensation and recording the residual error. At first there were few instrument specialists in the squadrons, and compensation was a tedious undertaking, requiring the better part of a day for a single airplane. Two or more compasses were installed on each bomber, and each machine had to be swung twice—once with bombs on board, once without, because dropping the bombs changed the compass deviation.[62]

Before each flight the navigator computed the flight plan. With the route drawn on a chart, he measured the course and distance. If there was a weather station at his base, he could get a forecast wind. If not, a short flight was made over the home base, to measure the wind at altitude. The navigator then combined the expected airspeed of his aircraft with the course and wind data, using a DR computer, and found for each leg the heading to fly and the groundspeed. The groundspeed applied to the distance gave the estimated time en route.[63]

Once in the air, the airplanes usually climbed to flight altitude over the home base. The bases were close to the lines, and it was unsafe to cross the lines at altitudes where small-arms fire was effective.

Further, by departing the base at cruising altitude, the navigator could fly his computed heading and check the accuracy of both heading and groundspeed when he passed the first ground beacon. For the Allied bombers based between Nancy and Belfort, there were usually two lighthouses between the base and the trenches, so that the course and speed could be well established before entering hostile territory. From the trenches to the target, the crews used any landmarks that were visible to verify their positions. For accurate dead reckoning, heading changes were kept to a minimum. Otherwise, when a position was established by map reading, the navigator would not know the heading he had flown and could not compute an accurate wind. The need for a constant heading sometimes conflicted with the desire to avoid antiaircraft fire and searchlights.[64]

A few crews of the French night bombing Squadron 101, flying Capronis from a base near Nancy, developed night dead reckoning to a high level of precision. They accomplished several bombing missions by flying a compass heading from their own beacons to the target, with few or no map-reading fixes in between. The most famous similar exploit by the British crews stationed in the same area was probably a flight by 216 Squadron on 29 June 1918. The target was a chemical factory in Mannheim. In the face of foul weather, all the airplanes turned back except for a single Handley Page. The crew flew the precomputed compass heading, spiraled down through the clouds upon the expiration of their expected time of arrival, and found Mannheim below. On the return trip, having had no check of the wind since departure, they did not dare repeat the procedure over the hills around Nancy but continued to fly westward until their fuel was nearly exhausted. Feeling certain that they were at least on the French side of the lines, they again descended through the clouds and landed without damage in a field 130 nm southwest of their airfield.[65]

The German crews attacking England at night always had to cross the Channel on compass headings. The Gotha crews, who made most of the raids on England, navigated by map reading from their bases near Ghent to the coast, where they got a definite departure fix from pyrotechnic signals. Once they reached England, they resumed map reading, often following the Thames to London. The Giants, operating from the same bases, were large enough to carry a well-arranged navigation cabin, occupied by the aircraft commander when he was not bombing, and radio. The Giants therefore enjoyed most of the navigational advantages of the Zeppelins.[66]

The night bombers often had to spend more time circling the target area, identifying landmarks on a large-scale map, than did day bombers. The Germans developed an acoustical fire-control device of such accuracy that Allied crews learned to shut off their engines near the target and glide the last few miles. This maneuver of course changed the airplane's speed, drift, and altitude, so that bombsight settings had to be computed very close to the release point and then would probably be inaccurate.[67] Night bombers also continued to operate individually, after day bombers had switched to formation tactics. Each night bomber therefore continued to require a competent navigator. In formations, the navigator of the lead airplane does all the work, and newly assigned navigators can gain experience in the other aircraft.

Between the World Wars airmen debated the effectiveness of day versus night bombardment, drawing on the experience of the First War. While the argument embraced all facets of both operations, navigational and bombing accuracy were of central concern. Viewed closely, the record of the First War seemed ambiguous. Many fliers insisted that navigation had not been unduly difficult, that the crews had generally found their targets. A U.S. Air Service officer attached to an Italian squadron in the Po Valley was definite: "From experience in flying over cities in the raid areas at night, I may safely say that there is no way of hiding a city or railway yards even in the darkest nights with no moon and a sky covered with clouds. The cities observed were lighted with the minimum number of blue lights, having the lighting units so covered by opaque shades so as to hide every source from above. The outline of the cities, parks and railway yards were nevertheless distinguishable . . . ." The aviators of the Royal Navy, drawing on their current experience flying from Dunkirk to bomb naval bases on the Channel coast and airfields near Ghent, had reported during the war that "no difficulty is found on clear nights, even when there is no moon, in locating an objective." Major-General Trenchard, commander of the Royal Flying Corps, agreed with the naval view in 1917, stressing that the level of training had to be higher for night operations than for day. Soon after the Armistice, however, he said: "Though night bombing is the safer, many mistakes are made at night in reaching the locality it has been decided to bomb."[68] In good weather, German airplanes had found London easily enough, but the weather was not always good. For smaller, less distinctive targets, there is no reason to believe that German airplanes achieved greater accuracy than German airships. The frequent errors made by the latter were well documented and have been described above, but the British were unable to track airplanes as successfully as they had airships. An airplane was harder to see, and they were more numerous than the airships—no more than a dozen airships had been over England at one time. Such records of airplane accuracy as were kept were based almost entirely on crew reports, and the British, in analyzing their own 1940 efforts, were to find how often self-delusion causes aviators to believe they are where they want to be. The day-night bombing controversy within the RAF led in 1924 to an examination of the World War I results. "An examination of the records of the I. A. F. [Independent Air Force] shows that the number of night attacks which failed to reach their objective was proportionately far in excess of the number of unsuccessful day attacks."[69]

The third branch of aviation in which navigation was essential was overwater patrol. The naval Zeppelins spent approximately as much effort reconnoitering the North Sea as they did bombing England. The British, dependent on oceanic commerce, relied heavily on aircraft to escort convoys and to search for submarines, as well as to keep tabs on the High Seas Fleet. The British used small, semirigid airships (which use gas pressure to maintain the shape of the bag, as opposed to the Zeppelin, which had an aluminum or plywood skeleton); land planes; seaplanes (which look like ordinary land planes, except that the wheels have been replaced by pontoons); and increasingly toward the end of the war when the submarine menace became acute, large flying boats (which have ship-like hulls that ride directly on the water, with smaller floats often suspended from the wings). The most advanced of the latter was the F. 2a, a twin-engine boat developed from the Curtiss Company's prewar *America,* but larger, with a redesigned hull and much more powerful engines. The F. 2a, in service the last 18 months of the war, had a combat range of about 500 nm.[70]

No instruments or techniques other than those already described were available for overwater navigation, but the emphasis on accuracy was obviously greater, in keeping with the penalties that would follow mistakes. Even over water, map reading was used whenever possible. The North Sea was dotted with lightships, the international character of which was respected by the belligerents, and patrol routes frequently used lightships as turning points. Perhaps the most famous patrol pattern of the war was the "spider web," an octagonal figure centered on the North Hinder lightship halfway between Harwich and the Hook of Holland. German submarines entering or returning from the Channel traversed the area on the surface, to make better speed and conserve their batteries. Both the submarines, and the flying boats hunting them, used the lightship as a landmark. Where no lightship was convenient, buoys were sometimes used to mark patrol routes.

Model of instrument board used in DH-4 in 1918.
USAF photo.

On patrols in sight of the shore, navigators took running and 45°–90° fixes (Figures 1 and 2) with a bearing compass.[71]

To some extent, the pilot flying over land could afford to ignore the compass. Naval aviators had no alternative but to steer by it for most of their time in the air, and much of the impetus for an improved aviation compass and better compensation techniques originated with the Admiralty. In conjunction with the compass, driftmeters were also essential over water, both for measuring groundspeed and maintaining course. The Sperry and Wimperis driftmeters, described earlier, were well adapted to overwater use, with the navigator sighting either on the wave crests or on a floating marker dropped for the purpose. With these instruments, the navigator kept a detailed account of his progress by dead reckoning, updating the DR position when possible by fixes from map reading, bearing compass, or radio.[72]

When larger aircraft were employed on patrol, they were able to carry radios and use the direction-finding nets. The same directional radio stations that served the Germans so poorly over England were much more useful to patrol aircraft, merely because the latter operated closer to the stations. Bearing errors are customarily referred to in degrees, but when translated into miles the error varies directly with the distance. For example, if a Zeppelin over England, working with a station on the German North Sea coast, received a bearing that was only 3° in error, that error would be equivalent to about 15 miles. Another aircraft, using the same erroneous bearing but only 60 nm from the station, would find the bearing only 3 miles in error. After a long patrol, when the dead reckoning was apt to be doubtful, the increasing accuracy of radio bearings was especially welcome to homeward bound crews.

In the early part of the war, British crews often underestimated the strength of the westerly wind over the North Sea. Several aircraft were forced down by fuel exhaustion, and the early patrol aircraft were not very seaworthy. Toward the end of the war, the British crews had better instruments and became more adept at navigation, and losses attributable to navigational error decreased markedly.[73]

After surveying the accomplishments of aircrews involved in bombardment and patrol, it would be easy to overestimate the average level of navigational competence among World War I aviators. Of the total, only a few flew the missions that demanded close attention to navigation, and only they could "be dignified by the name of aerial navigator."[74] The wartime advances in navigational instruments have been described; yet Major Harry Egerton Wimperis, who was closely associated with aeronautical research, titled his paper read to the Royal Aeronautical Society on 30 April 1919, "Air Navigation: The Most Important of the Unsolved Problems Relating to Aviation."[75] Actually much that was begun during the war did not bear fruit until after the Armistice. The RAF had increased the quality and quantity of the navigation training given the crews who were to fly the Handley Page V/1500s to Berlin, but the missions were not flown and the crews did not demonstrate the results of the new training. Research underway in driftmeters, directional radio, and artificial-horizon sextants resulted, in the years following the war, in a number of new instruments. All such advances were welcome as aviation turned its attention to the establishment of commercial airlines. To attract passengers or win mail contracts, airlines would have to operate on fixed schedules; and that implied an ability to navigate through bad weather and at night.[76]

OVERLAND AIRLINES AND RADIO NAVIGATION

During World War I, aviation grew like a hothouse plant. What had been embryonic in 1914 was full-blown by 1918: aircraft were numbered in the tens of thousands; individuals employed in the design, manufacture, maintenance, and operation of aircraft numbered hundreds of thousands. And like the hothouse plant suddenly taken out of its artificial atmosphere, aviation without the stimulus of war faced a painful adjustment. Many individuals and organizations were determined that aviation should not revert to its prewar status of curiosity and sport, but should find profitable employment in the world of commerce.

Many pilots wanted to continue flying. Companies that had converted from other manufacturing activities to the production of aircraft could reconvert, but many aircraft companies had not existed before the war. Their staffs and facilities could not easily be turned to completely new activities. Air forces were intimately concerned; facing constricted budgets for an indefinite period, they hoped to use civil aviation as an inexpensive reserve, exactly as navies relied on the merchant marine. In Great Britain, commercial aviation offered particular advantage to those who were concerned with the future of the Empire. The evolution of Empire into Commonwealth was already well advanced, in fact if not in name. The drift toward separatism might be retarded by more rapid communications between the various parts, something commercial aviation could provide.[1]

In the popular mind, the airplane was a dangerous, if exciting, gadget. "Gymnastics and acrobatics in the firmament," useful as they were in military aviation, would not attract air travelers, who would "have no desire to take part in a circus."[2] Before people were apt to entrust their goods—let alone themselves—to air transportation, the airplane's bad press had to be changed. Too long associated with accident and destruction, the airplane's potential for safe, economical, long-range transportation had to be demonstrated. From such motives the *Daily Mail* renewed its £10,000 prize for the first aviator to cross the North Atlantic in 72 consecutive hours. The governments of South Africa and Australia offered prizes for the first flight from

Handley Page 0/400 night bomber converted to 16-passenger airliner for the London–Paris route. The wings have been folded. The small propeller turned a generator for powering the radio equipment.
USAF photo.

Great Britain to their countries. Aircraft companies supplied the aircraft, governments provided varying amounts of assistance. Eager aviators were available. And within 17 months of the Armistice, in a burst of spectacular flights, they proved their coming of age. The United States Navy, committing a large portion of the Atlantic Fleet to the task, shepherded an NC (Navy Curtiss) flying boat from Long Island to England, via Newfoundland, the Azores, and Lisbon. John Alcock and Arthur Whitten Brown, both former RAF officers, flew a Vickers Vimy bomber from Newfoundland to Ireland nonstop. An RAF airship, the *R 34*, flew from Great Britain to New York City and returned. Two Australians, Ross and Keith Smith, flew a Vimy from England to their home; and two South Africans, Pierre van Ryneveld and C. J. O. Brand, after crashing twice, reached Capetown in their third airplane.

On the Australian flight, the Smith brothers demonstrated what could be done with the battlefield navigation they had learned during the war. Without radio, sextant, or even turn indicator, they planned to rely on map reading as much as possible. They would trust an aperiodic compass to cross the unavoidable stretches of open sea, correcting the course by observations through a Wimperis driftmeter. And only twice on the entire trip did the scheme fail. On the first leg over France, the clouds were so low that they could not stay in visual contact with the ground but had to climb on top. They flew a compass heading for three hours, found a hole, descended, and were much reassured to find that they were almost exactly on course. Again, crossing the mountains from Rangoon to Bangkok, they found clouds filling the passes. The airplane ceiling of 11,000 feet did not allow them to top the clouds, so they flew through the clouds. Without blind-flying instruments they spent a difficult hour and then decided that the danger of getting into a spin exceeded that of hitting a peak, so the pilot shut off the engines and glided down as slowly as possible. "As we had not hit anything by the time we reached 4,000 feet, I concluded that the range had been crossed." In areas where maps were blank, they followed coastlines, rivers, railroad and telegraph lines. They crossed Australia in short stages, asking directions of cattlemen at each stop.[3] The Smiths won the prize, and Keith, the navigator, used his primitive navigational techniques for all they were worth. Two years later, when the brothers were planning a round-the-world flight, Keith tacitly admitted that something more was needed: he enrolled in a navigation course at RAF Station Biggin Hill. They were not, however, to make the world flight. Ross Smith died in a crash while testing their new airplane.[4]

Most of the crews who attempted the African and North Atlantic flights in 1919 and 1920 relied on more sophisticated means of navigation—radio and celestial. The largest airplane entered in the North Atlantic race, a Handley Page V/1500, carried the airborne direction-finding coils that had been developed specifically for that aircraft. The American NC flying boats and the airship *R 34* were equipped with similar direction-finding equipment.[5] The crews of these aircraft were therefore able to take bearings on coastal radio stations as well as on ships along their routes. The smaller airplanes could carry only communications radio; those crews intended to request bearings from the same sources. The wartime development of the radio equipment has been described above (pp. 84–93). Experimentation with air sextants and rapid reduction methods had also begun during the war, but celestial navigation had been used only slightly, if at all, in airplanes.

In the United States, the Division of Science and Research of the Bureau of Aircraft Production sponsored an investigation into the possibilities of celestial navigation for aircraft at Langley Field, Virginia. From August 1918 to January 1919, an impressive array of talent and resources under the direction of Henry Norris Russell was devoted to the study of the same problems that European balloonists had grappled with before the war: how to measure the height of a celestial body and reduce the measurement to a position while in the air. Russell, of the Princeton Observatory, was assisted by James Percy Ault, a marine navigator from the Department of Terrestrial Magnetism of the Carnegie Institution. Russell's pilot was Captain D. L. Webster, who on demobilization became a professor at the Massachusetts Institute of Technology. When Professor R. W. Willson of Harvard heard of the work, he took a bubble attachment of his own invention to Langley for evaluation. Russell had the use of the Science and Research Laboratory at Langley, plus Army and Navy aircraft at Langley and Norfolk. After six months of flight testing, in which more than a thousand observations were taken from nine different types of airplanes both day and night, Russell reported that celestial navigation was indeed feasible in the air. He showed that when flying at low altitudes during the day, over water or flat land, the marine sextant could be used to measure the height of a body above the natural horizon without undue error. He investigated the use of cloud and haze horizons as reference lines, but found them unreliable. Finally he proved that the marine sextant could be used with either a separate artificial horizon (a flat mirror mounted on a pendulum, the oscillations of which were damped by suspension in oil) or an air-bubble attachment. With either arrangement, he recommended two procedures for the sake of accuracy: the pilot, who had to be highly skilled, must be especially careful to steer a straight course while the observations were being taken; and the observer must take at least six observations of the same body, discard those widely at variance with the others, and average the remainder. When those conditions were met and the air was reasonably smooth, Russell achieved an average accuracy of 6 miles.[6]

The investigation encompassed methods of reduction as well as sextants. Russell recommended using St. Hilaire's intercept method of plotting. Precomputation on the ground reduced the time required in the air and minimized the danger of arithmetical error, but all computations could be made in the air when required. For the latter, he favored Aquino's Tables[7] or a new circular slide rule designed by Charles Lane Poor. The slide rule was "a little large for convenient use in a small airplane [it weighed 9 pounds] but a smaller one could easily be made."[8] With precomputation, observing, reducing, and plotting a single line of position required no more than ten minutes; two lines, to provide a fix, required less than 20 minutes.[9]

The tenor of Russell's report was optimistic—too much so, apparently, for his assistant, J. P. Ault of the Carnegie Institution. Ault, when asked by some Army officers "as to what he thought would be the successful method for navigating aircraft in the future," replied, "by the use of radio." Celestial navigation was "practical and feasible [but] too slow and uncertain to be relied upon for future aerial development."[10]

Contemporary with Russell's investigations at Langley, a small group of American naval officers in Washington were planning a trans-Atlantic flight. The Secretary of the Navy approved the project in February 1919, after Russell's study was completed but before the results were published. Lieutenant Commander Richard E. Byrd, who had long interested himself in the problems of overwater navigation, was charged with developing the navigational instruments for the flight. He designed a bubble sextant similar to the Willson model, which Russell had tested.[11]

At the same time a bubble sextant was under development in England. A patent awarded to Lionel Barton Booth in October 1919 became the basis of a series of sextants produced by the Royal Aircraft Establishment, Farnborough. The R.A.E. bubble sextant, Mark I, was used by one of the unsuccessful crews who attempted the African flight in early 1920. The German rights to the Booth patent were purchased by the Plath Company, which produced the instrument under the trade name SOLD-Sextant. The French placed their hopes in the development of a gyroscopic sextant, but the instrument remained inconvenient to handle and difficult to maintain.[12]

The bubble sextants of 1919 and 1920 were remarkably similar, differing only in details such as weight and balance, illumination, location of the handle, and method of reading the scales. The optical systems were designed so that the bubble and the celestial body, which had to be viewed together, would appear at the same optical distance from the eye and would move in the same direction, at about the same speed, when the sextant was jostled. The instruments seem to have given similar results. The average error of the R.A.E. Mark I was less than 10 miles on the African flight. A British report of May 1921, comparing the R.A.E. Mark IIa with the American Willson sextant, found that "much the same accuracy is obtainable by the two instruments, the mean error of the latter being slightly less."[13] All of the sextants, either in their original models or in early modifications, permitted the navigator to remove the bubble from the field of view and sight on the natural horizon when conditions permitted, because the natural horizon, when available, always gave superior results. The British agreed with Russell that cloud and haze horizons should not be used, because the altitude of the top of the cloud layer usually

Bureau of Standards bubble sextant.
Weems & Plath photo.

could not be measured precisely. The navigator would therefore be unable to correct for dip, the error caused by the cloud horizon being below the observer's horizontal plane.[14]

One British sextant, which appeared in 1919, was designed to overcome the unknown dip error. The Baker air sextant added a second horizon prism, pointing in the direction opposite to the first, so that the navigator could see at one time the body, the horizon directly beneath the body, and the opposite horizon. By placing the body precisely between the two horizons in the optics, the observer effectively bisected the arc between the two opposite horizons and thus defined the vertical. The instrument was for use when flying on top of a cloud or haze layer. The basic assumption, that the cloud layer would be the same height in both places, could easily be untrue; in any event, the navigator would not know. Further, the instrument could not be used at night, and the navigator too often found the section of the horizon directly beneath the body blocked by part of the aircraft. Chances were twice as great that one of the two sections of the horizon required by the Baker sextant would be blocked from his view. Baker's attempt to avoid the use of a bubble level was soon recognized as a failure. The instrument's one chance for fame was thwarted. Commander Mackenzie Grieve, Harry Hawker's navigator in their unsuccessful Atlantic attempt, intended to use a Baker sextant, but it was broken in transit, and he bought a marine sextant in Newfoundland.[15]

Of the successful trans-Atlantic flights, that of Alcock and Brown commands particular admiration for, among other reasons, the excellence of the navigation. Brown had been an observer in the war, flying artillery spotting and photographic reconnaissance missions. He was shot down, injured, and captured. As a prisoner of war, he became interested in aerial navigation. His plans for the flight were well laid, including the best instruments available in the summer of 1919. He practiced celestial navigation on the bridge of the ship taking the airplane and crew to Newfoundland, swung the compasses once the airplane was assembled, and checked the radio with the ground station on local test flights. No bubble sextant was available, so he carried a marine sextant and an Abney spirit level, a surveying instrument, to use as an artificial horizon. Byrd traveled as far as Newfoundland with the NC flying boats and offered to lend Brown a bubble sextant, but it did not arrive from the States in time.[16]

On the flight, the small propeller that powered the radio generator soon snapped off. Alcock and Brown were plagued by difficult weather for most of the crossing, so that the driftmeter and the sextant could be used only occasionally. Following their noon takeoff, Brown was able to establish their position twice before sunset with drift observations and sunlines. The drift, added to the heading, provided the course. Position lines from the setting sun plotted nearly at right angles to the course line, so that the aircraft's position was fixed at their intersection. In the middle of the night they were on top of the clouds long enough for Brown to plot a two-star fix and alter heading on the basis of the new position. Toward morning Brown had to alternate between navigating by dead reckoning and crawling out onto the wings to chip ice off the engine air intakes. With nothing more than dead reckoning from that point, they would have hit Ireland; but a short observation of both sun and sea at sunrise enabled Brown to get a fourth position, from which he altered heading for Galway. When they crossed the Irish coast, they were almost precisely on the new course.[17]

Brown reduced the celestial observations with the aid of a Baker navigating machine, a device similar to Brill's plotting machine, which dated from 1909 (see above, p. 35). Brown's chart of the route, a Mercator, was placed in the Baker machine, and one of two transparent charts was positioned over the Mercator. One transparent chart carried a set of Sumner circles for use with sun observations; the other carried circles for six stars that would be in suitable positions during the night. In his celestial work, Brown thus avoided the use of tables and lengthy computations in the air, but he chose to work the dead reckoning with marine traverse tables (see above, p. 10). Using the sextant in the open cockpit was not easy. He had to twist himself into various positions to align the sextant with the celestial body and brace himself as well as possible to hold the instrument steady in the airstream. At night he also had to manipulate a flashlight.[18]

The navigational task facing the American NC flying-boat crews seemed much simpler, because of the vast organization devoted to the effort. Three aircraft attempted the flight, following a route chosen to reduce the length of each stage: Long Island, Newfoundland, the Azores, Lisbon, England. Destroyers were spaced every 50 miles along the 3,800 mile route, to provide navigational assistance and to rescue downed airmen. By using some ships more than once, a total of 68 sufficed. During the day the destroyers made smoke; at night they lit up searchlights and fired star shells. Further, the flying boats were large enough to carry every available navigational aid.[19] In the view of an Englishman who did not consider the weather, "the Americans had nothing to do but navigate by sight. . . . the navigational problem was reduced to the last terms of ease"; and an American participant agreed that it should have been "as easy as 'walking down Broadway' "—except for the rain and fog encountered during the night of the Newfoundland–Azores leg.[20] Direction-finding radio was the only navigation aid that could possibly operate inside clouds in 1919, and it failed all three aircraft during the night. Thunderstorms along the route produced severe static and the magnetos of the engines also produced radio noise. From magneto interference alone, one flying boat could take no radio bearings from sources more than 10 miles away. The crews of all three boats were uncertain of their positions approaching the Azores. Landing in high seas, one boat sank and another was damaged. Only the third, which landed in the shelter of one of the islands, could continue the flight. But directional radio, performing acceptably in clear weather, helped keep the last boat from going astray on the next leg to Lisbon. The steering compass was jarred loose on takeoff, and the pilot took up a course 8° too far right. As the destroyers appeared farther and farther to the left, the navigator used the radio to home on the fourth ship, and later discovered the compass error. The NC navigators used the bubble sextants only occasionally.[21]

Two more flights of the early 1920s demonstrated the possibilities of aerial navigation. The RAF conducted an exhaustive test of its navigation instruments, excepting radio, in August 1920. A well-qualified crew, including some R.A.E. instrument designers, flew a Handley Page 0/400 around Britain, over 1,000 miles in 11 stages.

The compasses and driftmeters gave excellent results in dead reckoning: at the end of legs varying from 25 to 200 miles, the DR error varied from 0 to 9 miles, an average of 3.3 miles. The celestial results, using an early model of the R.A.E. bubble sextant, were less impressive. The error in 16 sunlines ranged from 2 to 43 miles, averaging 14 miles. The sights were reduced with another new instrument: a slide rule devised by Captain L. C. Bygrave of the Air Ministry Laboratory. The difficulty with all previous mechanical and graphical reduction devices had been that, to secure acceptable accuracy, the instrument had to be so large that its use in a cockpit was difficult or impossible. By using a series of equations involving only cosines and tangents and wrapping the two required scales in a spiral around telescoping cylinders, he packed 56 feet of scale onto an instrument only 14 inches long, 3 inches in diameter, and weighing less than 2 pounds. He provided instructions on the instrument, arranged in a series of steps, so that a person who knew little of the theory but who had mastered the drill could work out an observation in about four minutes of time, to an accuracy of 1 or 2 miles.[22] As would be true for many years, the weakest link in celestial navigation for aircraft remained the sextant. A Portuguese naval officer, Gago Coutinho, carried the instrument one step further toward perfection.

Since 1919 Coutinho had considered flying from Portugal to Brazil on the centennial of the political separation of those countries. The route did not follow a shipping lane, and there were no powerful radio stations on shore, so he did not plan on using radio. The Portuguese could provide only a small, single-engine seaplane, incapable of making the journey nonstop; refueling would be necessary at four islands. The most difficult leg would be the thousand miles from the Cape Verdes to St. Paul's Rocks, a speck on the equator 600 miles northeast of Natal. The volcanic rocks covered an area of 650 square yards, with a maximum elevation of 30 feet; they would not be easy to find.

Coutinho painted drift marks on the tail of the seaplane, but he placed his greatest reliance in celestial navigation. He designed a bubble sextant and tested it on flights in 1921. It differed from those previously described in that, instead of a single, spherical bubble-chamber, it had two tubular chambers, perpendicular to each other, much like a carpenter's level. The pilot was expected to steer toward the celestial body during the observation, and the longitudinal level was therefore of primary importance. Because the radius of curvature of the spirit chambers was larger than in the spherical chambers used in the sextants described previously, the bubbles in the Coutinho instrument reacted more quickly to changes in the vertical—and to

MOST PROBABLE POSITION
112

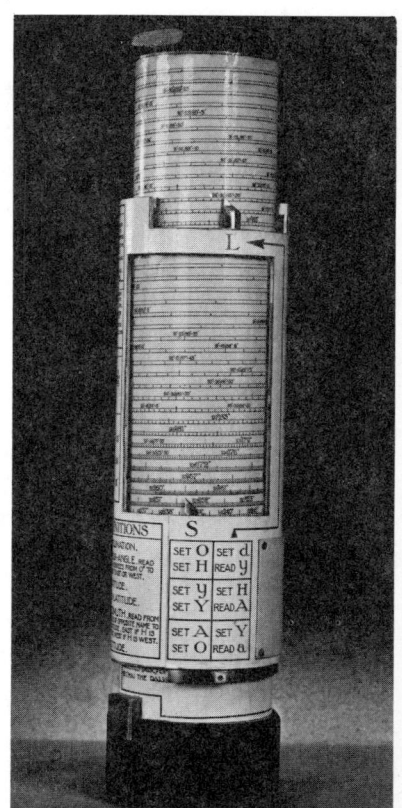

Bygrave slide rule.
*Henry Hughes & Son photo,
courtesy of Capt. Weems.*

Drift lines painted on the rear fuselage and horizontal stabilizer of Fokker Tri-Motor.
USAF photo.

extraneous accelerations—than did the other artificial-horizon bubbles. This meant that the navigator had to use more skill in observing, but more accuracy was possible in smooth air, with a careful pilot. In April 1922 both pilot and navigator demonstrated the requisite skill: Coutinho plotted 40 sunlines in the 11-hour, 20-minute flight from the Cape Verdes and found St. Paul's Rocks. Because they were never forced to fly above clouds, all celestial sights were taken while flying very low over the ocean and using the natural horizon. Coutinho's experimentation with more sensitive bubbles would influence later sextant designs, and their 1922 achievement stands as one of the most remarkable feats of aerial navigation. The seaplane capsized on landing, but the crew ultimately reached Brazil in their third aircraft.[23]

The postwar demonstration flights proved that existing aircraft were capable of long flights. But the number of crashes involved, and the complete lack of commercial cargo carried (the airplanes usually being dangerously overloaded to have enough fuel for the flight) did not augur well for long-distance airlines. The navigators had usually found their way safely, and their newest tool, the bubble sextant, was shown to be capable of remarkable accuracy when handled by highly trained individuals.[24] Radio had not done so well, but advocates were not lacking to point out that enormous improvements were just over the horizon. The navigation used on the first flight described above, that of the brothers Smith to Australia, had seemed old-fashioned at the time; Keith Smith had relied almost solely on map reading. The 1920s were to show, however, that navigation on the world's first commercial airlines would resemble Keith Smith's more closely than Gago Coutinho's.

If the demonstration flights had not demonstrated extreme reliability, many commentators continued to insist that such was the first prerequisite for commercial aviation. By reliability they meant safety, of course; but they also emphasized regularity. If flights did not arrive and depart on schedule, some tourists and "those who are apostles of aerial travel" might fly; but the businessman had to plan on arriving at a given time.[25] For mail contracts, too, regular operations were essential. To compete with express trains, airlines had to fly at night; otherwise the night train, whose speed was about half that of the contemporary airliner, would make up the distance lost to the airplane during the day. To exploit the potential of high speed, the airliner also had to fly at the altitude where favorable winds were found, which would sometimes be above a solid overcast. Indeed, smoother air was often found above clouds—an important consideration, since airsick passengers were not apt to become steady customers. Flying beneath the clouds, following a railroad for example, was therefore unwise for a number of reasons: a longer route, less favorable winds, a bumpy journey—and when the clouds were very low, a greater possibility of flying into the ground. All of these arguments implied that the navigation of airliners had to depart from the common practice of most World War I aviators. Some pilots would object to the change, but if they were told "the facts, preferably expressed in £.s.d.," they might accept the necessity of more advanced navigational techniques.[26]

The methods of navigation, like every other aspect of commercial aviation, required international agreement and cooperation. In Europe, even the earliest airliners were capable of flying beyond national boundaries; and it was easy to foresee that the range of aircraft would continue to increase, so that in time a single flight could easily cross several nations. Assuming that international agreement would be easiest to attain in conjunction with the peace negotiations, the British prepared a draft proposal that resulted in the *International Air Navigation Convention* of 1919 and the International Commission for Air Navigation, a continuing organization.[27] Attempting to foresee the future needs of commercial aviation led to many regulations that would not be pertinent to the 1920s.

The *Convention* required that aircraft in public transport carrying more than ten passengers a distance of 272 nm over land, or 108 nm over water, or at night have a navigator on board. The signatory nations undertook to examine and certify aerial navigators, issuing a series of certificates permitting an individual to serve as navigator under those various conditions.[28] But the first airliners would not seat ten passengers, did not fly at night, and usually did not fly routes as long as those specified above. When aircraft of larger capacity and longer range came into use, the airline companies had learned the economic necessity of the minimum flight-crew. Thus, although navigator certificates and syllabuses were prepared and examinations advertised, there were few applicants, because there were no jobs. In 1925 the Commission altered the rules to allow one of two pilots on board to act as navigator, if properly certified. Even so, at the end of 1926 there were only ten navigator licenses current in Great Britain, as against 286 pilot licenses. A book describing the opportunities of flying careers in 1934 noted: "Up to the present, no professional air navigator who is not also a pilot exists so far as the writer can ascertain. But it is likely that in the future positions for such specialists will be found," when larger commercial aircraft come into use.[29]

In addition to specifying an unrealistic array of aircrew positions, the 1919 *Convention* also required a series of logs to be maintained on each commercial aircraft. As it became apparent that the various logs were unduly repetitious and that the pilot was filling in most of them after landing, the several logs were combined into one. The bookkeeping was further simplified in 1927, when the 13th session of the Commission resolved that "the sections of the Journey Log Book headed 'Navigation' need not be entered up" and other requirements were deleted.[30]

The delegates who drew up the *Convention* contented themselves with specifying the qualifications of aircrew members and the forms they should keep and did not attempt to prescribe the routes airliners should follow. In practice, the pilots continued to follow the roads and railroads. The 1st International Congress of Aerial Navigation, meeting in Paris in November 1921, was advised to consider "the danger resulting from this practice. On a given aerial route, all the pilots would follow, in cloudy weather, exactly the same landmarks, thus creating great risk of collisions." The same month a paper presented to the Royal Aeronautical Society urged that, during bad weather, aircraft flying opposite directions be at different heights.[31] Imaginative prediction of tragedy does not often move men to action, however, and the foreseen event followed five months later. A northbound British airliner collided head-on with a southbound French airliner. The ceiling was low, visibility poor, and the pilots were apparently following the same road. The accident was spectacular enough to produce action. The company managers and the pilots talked; the solution seemed to lie in establishing traffic rules and routes, but the pilots had favorite landmarks and could not agree on a common route. Ultimately the British, French, Belgian, and Dutch governments established a series of official routes and required that pilots stay to the right of the center line when using them.[32]

The New World was less hasty in regulating commercial air travel, whether domestic or international. Canada conformed with the provisions of the *International Convention* by establishing a procedure for licensing pilots and "others" engaged in flying; but an official report of 1924 noted that "air navigators' certificates ... in the present state of development of aviation have not been found necessary." The earliest U.S. domestic regulations, effective at the end of 1926, did not recognize the crew position of navigator. Transport pilots had to take a "theoretical examination in the fundamentals of meteorology and air navigation."[33] The flight test required a flight of 100 miles and three landings, but it could be waived if the individual had flown solo at least 100 miles within the previous year. The *Habana Convention on Commercial Aviation,* which the U.S. Senate ratified in February 1931, dealt with international flights in the Western Hemisphere, but it set no requirements concerning navigation.[34]

The 1919 *International Convention* and the subsequent meetings of the Commission took up the matter of aerial charts where it had been dropped by the international aviation organizations before the war (see above, pp. 19–22). A bulky Annex F to the original *Convention* contained detailed specifications for preparing two series of maps to cover the world. The Annex was extensively amended in 1923; in 1932 a third and fourth series were added. The delegates were thorough—rivers were to be distinguished as to: perennial and surveyed, perennial and unsurveyed, nonperennial and surveyed, nonperennial and unsurveyed; railroads as to: single or multiple tracks (number given by Roman numeral), in use, under construction, abandoned, "light railways," tramways—because the production of the maps was divided among the signatory nations. If the product was to be uniform, the different agencies had to follow common rules. But attempting to comply with the changes must have frustrated the cartographers, and any maps produced were in imminent danger of obsolescence. Britain and France made good progress with their sections of Europe and of the colonial empires, but the task was beyond the resources of many countries and no single series was complete by World War II.[35]

The United States went its own way in preparing aerial maps, and the entire nation was mapped by December 1936. The need for air maps had become glaringly apparent with the beginning of the airmail flights in the last months of World War I, first by the Army, soon by the Post Office. The problems encountered in prewar Europe were met again in postwar America. Pilots used whatever maps they could find, including pages torn from atlases. The Europeans had explored the photomosaic and dismissed it as unsuitable for a flight chart (above, pp. 21–22). Nevertheless the U.S. Army Air Service began an ambitious program to photograph all air routes (from both directions, so that a pilot could see the exact view he would encounter) and to use the photomaps for orientation. Commercial companies helped fill the need for charts, the most useful offering being Rand McNally's *Air Trails Maps,* published in 1928. These were standard state maps overprinted with aeronautical information. Because the maps were arranged one state per sheet, the scale varied widely, but *Air Trails Maps* were the best series providing national coverage for eight years and were therefore widely used.[36]

When agencies of the U.S. government undertook the production of aerial maps, they chose the strip form as the most easily handled in a small cockpit. By printing a narrow chart connecting two cities, they saved the aviator the problem of cutting and pasting larger sheets. The Air Service began issuing strip charts of the principal air routes in 1923. From 1926 the Navy's Hydrographic Office produced a series of strip charts covering the coasts of North and Central America. Following the passage of the Air Commerce Act of 1926, which vested control of civil aviation in the Commerce Department, the Coast and Geodetic Survey began producing yet another series. The nation was being criss-crossed with strip maps prepared by various agencies, with some routes being mapped more than once while certain areas remained unmapped. A committee representing the interested agencies of the federal government considered the matter of aerial charts throughout 1929. Statistics showed more flying was being done off the airways than on them, and no proliferation of strip maps was apt to supply the needs of private aviators who usually flew off the airways. For military purposes, too, the nation needed a systematic, inclusive air map. The committee therefore recommended that the Commerce Department undertake such a series. The new chart, called the Sectional Aeronautical Chart, is a conic projection on a scale of 1:500,000. Most of the charts cover 2° of latitude and 6° of longitude on sheets about 20 by 40 inches. The exceptions are along the coasts, where rigid application of the formula would result in charts showing little other than ocean. The first of the charts appeared in 1931. They were produced at the rate of one every two months until late 1934, when an infusion of Public Works Administration funds quadrupled the rate. The last of the 87 appeared in 1936, by which date the first had been corrected and reissued ten times.[37] The series remains of major importance for private fliers and has certain military applications.

Along with the production of maps, efforts to make the surface of the earth more distinctive for map-reading aviators resumed on both sides of the Atlantic after the war. During the war, aerial routes had been marked with star shells and captive balloons, each of the latter riding on the end of a "horrible wire passing up through the clouds." Both methods helped overcome the effects of ground fog and low clouds; but without the spur of military necessity, such procedures became less attractive to aviators.[38] The 1919 *Convention* dealt with less hazardous artificial landmarks, again picking up the debate where it had been interrupted by the war (above, pp. 56–57). The same arguments were advanced in favor of extensive systems to be erected and maintained at government expense, and many ground marks were installed. Civic pride was successfully stimulated: some chambers of commerce were determined that their cities be adequately advertised to the air-age traveler. Boy Scouts whitewashed rocks for airway signs; the General Electric Company helpfully provided plans for rooftop signs (which would of course be illuminated).[39] Public-spirited railroads aided their future rivals by allowing signs to be painted on station roofs. Given the widespread habit of following railroads (the "iron compass"), the station roof was in great demand as a site for signs.[40] In the end none of the extensive schemes[41] was completed for two reasons: anything approaching complete coverage of the countryside was prohibitively expensive, and the volunteer programs produced only a scattering of ground markers. An aviator discounted the results achieved by 1939: "These signs are just like cops. I've never seen one when I needed it most." One case of ground marking was uniquely successful. The RAF, in establishing an airmail route from Egypt to Bagdad after World War I, plowed a ditch across the sections of desert most lacking in distinctive landmarks.[42]

Night flying had ceased to be a novelty by the end of the war, and many kinds of lights had been used as aerial beacons. Some wartime expedients, such as pyrotechnics, were not applicable to peacetime aviation. Airport lighting methods had depended on the military's prodigal use of manpower; no civil concern could afford it. Route beacons during the war had been relatively weak; when the lights came on again, the wartime beacons were no longer visible.[43] The victorious nations set to work energetically to overcome the lead established by the Germans before the war; and throughout the interwar period, more and brighter lights were placed in service along air routes all over the world.

As with photomosaics and overly ambitious ground-marking schemes, many lessons learned before the war apparently had been forgotten, at least by the Americans, who experimented with vertical beams and the use of colored lenses in aerial beacons. Both had been tried and discarded in prewar Germany (above, p. 57). The British Air Ministry tapped their national maritime experience (represented by Trinity House, a semi-public corporation that has maintained English coastal lights and buoys since the reign of Henry VIII), designed and installed a variety of experimental lights, and compared their performance under varying weather conditions. The French proceeded similarly so that by 1922 the London–Paris route had lights installed and a test flight was flown.[44] The beacon design generally favored, after the initial experiments, featured a main incandescent light, uncolored, and rotating, to give the effect of flashing without being turned on and off in rapid succession. Smaller colored lights were added to identify the particular beacon or to mark the direction of the airway. The French won the distinction of having the brightest beacon. The French General Staff started work on a 1,000,000-candlepower beacon near Dijon, to guide bombers back to their bases, in 1918. Interrupted by the Armistice, the work was completed in 1925

CAA beacon displayed at National Aeronautic Forum and Convention, Bolling Field, D.C., May 1940.
USAF photo.

for civil aviation on the Paris–Marseilles–North Africa route. The beacon used eight lenses, each about 5 feet in diameter.[45]

American industry never matched the French giant, but developed a series of efficient beacons that were competitive internationally. Australia chose an American product, the 24-inch Sperry electric beacon, for 900 miles of the Perth–Adelaide route in 1929. The United States Post Office first used acetylene beacons, then rotating electric beacons of 18-inch diameter, later 24-inch diameter, along the airmail routes in the mid-1920s. By 1933 the standard equipment was a 36-inch double-ended beacon, turning at three r.p.m., giving six flashes every minute.[46] For installation in isolated areas, generators were provided, and the lights were designed to operate without attention for extended periods, even changing burned-out bulbs automatically. Vandalism seems to have been a problem. An RAF pilot, flying the mail from Cairo to Bagdad, recognized the need for beacons for night operations but speculated that "the Bedou might treat them as Aunt Sallies and practise their rifles on them." The U.S. was not immune. Beacons installed on the Chicago–Cheyenne route in 1923 carried this placard: "Pilots depend on this beacon for their safety. Property of U.S. Air Mail Service, Post Office Department."[47]

No matter how efficient beacons became, nor how many were installed, in really thick weather they were useless. An airmail pilot is supposed to have quipped: "If you can see lights it shows you don't want them; if you can't it shows you do." For navigation in and above clouds, directional radio was obviously necessary. To most of those who spoke on the subject, this wartime development seemed a panacea.[48]

Wartime developments had resulted in several radio direction finders: the Telefunken directional transmitter, the ground directional receiver operating singly or in a net, the coil wrapped around the airplane wing and used for homing, and the airborne rotatable coils that could be used to measure the bearing of a transmitter without turning the aircraft. Regulatory bodies, airline officials, and equipment manufacturers had to decide which system was best suited for commercial aviation. Navigators would have preferred having the equipment in the aircraft, so that they could measure the bearings themselves; but navigators were not to be employed by civil carriers. A pilot would find plotting a bearing difficult or impossible. Placing the direction-finding equipment on the ground reduced the weight the airplane had to carry; ordinary communication radio, which airliners would be required to have, was sufficient to request and receive bearings from ground stations. Airline officials also noted that the expense of equipment at the ground stations would be borne by the government. Direction finding required complicated circuits; trained maintenance men could monitor ground installations and correct malfunctions quickly. All these arguments led to placing the direction-finding equipment at ground stations.[49] The Europeans quickly converted their ground directional receivers to civilian uses; the Americans, after a few years of experimentation, developed ground directional transmitters that could be used by ordinary receivers, with little or no additional equipment in the airplane.

As airlines were established linking the capitals of northwestern Europe, direction-finding stations were established to serve them. By 1925 the British, French, Belgians, and Dutch had integrated their systems, netting ground stations without regard to political boundaries. To overcome the language difficulty, a vocabulary of standard phrases, each identified by a number, was published. Thereafter if a pilot could count in the language used by the net control stations, he could get either a bearing to an airfield or his position. Unless too many aircraft needed assistance simultaneously, a pilot could expect to receive his position within a minute and a half after requesting it.[50] The equipment was continually improved, and the errors that had degraded the bearings supplied to the Zeppelins during the war were identified and, when possible, eliminated by redesign.

One such error, called night effect, referred to the odd behavior of radio signals during nighttime, especially at sunrise and sunset. Significant and changing errors were introduced into bearings at such times, if the receiver was more than about 40 miles from the transmitter. By 1918 English researchers suspected that somehow horizontal antennas were related to night effect. By 1921 the theorists supplied the explanation: during the night the lower layer of the ionosphere, about 25 miles above the earth, reflects radio waves that strike it at slight angles (such as waves emitted from a horizontal antenna). During the day, sunshine breaks up the reflecting layer, but at night waves are returned to the earth where they are picked up with other waves that have traveled the direct route from transmitter to receiver. If the two waves arrive in phase, they reinforce each other and produce a stronger signal; but just as often they arrive out of phase and cancel each other. Further, reflection often bends the waves not only vertically but horizontally. Once understood, night effect could be eliminated by using only vertical antennas for direction finding. A ground station using four vertical antennas, instead of crossed loops, was patented in England in 1919 but nothing further was done to develop the equipment until the late 1920s. Engineering the new antennas proved troublesome because the vertical antennas, spaced up to 400 feet apart, had to be fed by horizontal lines. By shielding the horizontal lines, coating them with lead or copper and burying them, a design was achieved that permitted no radiation from horizontal members. By 1932 the new antennas were being installed in both the United States and Great Britain. At the same time short, vertical antennas were placed on airplanes, replacing the trailing wire for direction-finding purposes.[51]

A second error is called coastline effect. Radio waves crossing from land to water, or vice versa, are bent much like light rays passing from air to water. The error cannot be prevented, but by 1921 aviators were being warned that coastline effect caused bearings to be inaccurate in certain areas.[52]

A third error was identified: irregular terrain reflects radio waves. The British call this land effect; it did not cause them much trouble.[53] When directional-radio stations were constructed in the Rocky Mountain West in the 1930s, terrain reflection was found to be severe. In the United States the error is called mountain effect; attempts to correct it will be described below.

A fourth error in directional radio is 180° ambiguity, caused by the inability of the coil to differentiate between signals coming from opposite directions. This fault could be overcome by plotting bearings from two or more stations; but the ability to determine the correct direction of a single bearing (called determining the sense of the bearing) was needed. A simple circuit combining a loop with a straight-wire antenna provided the answer. A loop antenna receives maximum signals when in line with the direction of the transmitter, minimum signals when perpendicular to the direction of the transmitter. If the loop is in line, receiving a maximum signal, and then rotated 180°, the signal induced in the antenna will again be maximum, but of opposite polarity (a positive voltage in one case, negative in the other). A straight-wire antenna is nondirectional and receives the same signal from all directions. In a sensing circuit, the straight-wire antenna is designed to receive a signal of the same strength as the loop in its maximum position. When the outputs of the two antennas are combined, the signals will be additive at one of the loop's maxima, but at the other they will cancel, producing a sharp minimum signal (Figure 12). This circuit, worked out for ground and shipboard direction finders in the 1920s, was included in airborne direction finders when they became common in the 1930s.[54]

In spite of the many improvements made between the wars, the direction-finding system in use on European routes in 1939 was fundamentally the same as that employed during World War I. The United States, however, had taken a different path. At the end of the war, the governmental agencies concerned with aviation were busy

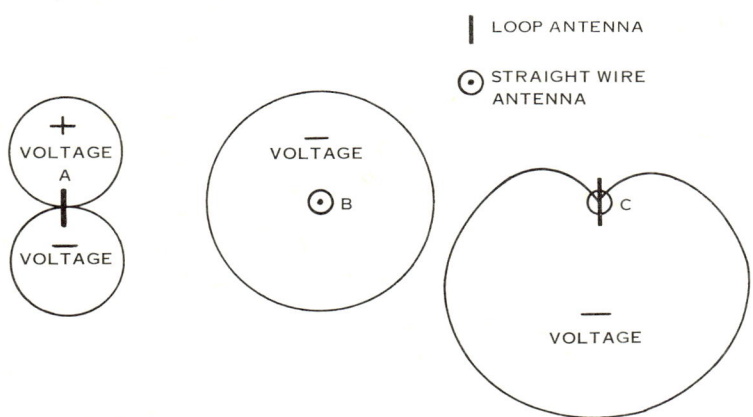

Fig. 12. Removing the 180° ambiguity from a radio bearing. Antenna A is a loop, seen from above. The distance of the curved line from the antenna represents the relative signal strength received from all directions. For A, the strongest signals are received in the line of the antenna, north and south. B is a straight, vertical wire; it receives equal signals from all directions. C combines A and B. The former maximum signal from the north has been canceled entirely, so that a sharp null results.

copying, modifying, and experimenting with the radio direction-finding equipment that had originated in Europe. The rotatable coils installed in the NC flying boats for the Atlantic flight were similar to those designed by the RAF for their anticipated Berlin raids.[55] During the period that the U.S. Post Office Department operated the airmail service, from 1918 to 1925, that Department cooperated with the Army, Navy, and Bureau of Standards to experiment with both rotatable coils and fixed coils wrapped around the airplane wings.[56] At the same time, the Bureau of Standards, at the request of the Army Air Service, developed a third radio navigational aid: a directional radio beacon that projected a beam which a pilot could follow to or from the beacon.

The principle of the beacon, or radio range as it was later called, had been patented in Germany in 1907. German efforts to develop the idea into practicable equipment had been only partially successful during the war. The range used two loop antennas, crossed, transmitting different Morse characters on the same frequency and at equal power. Each directional antenna transmitted its strongest signal along the line of the antenna. Halfway between the maximum signal lines the two signals were equal (Figure 13). This line of equal strength,

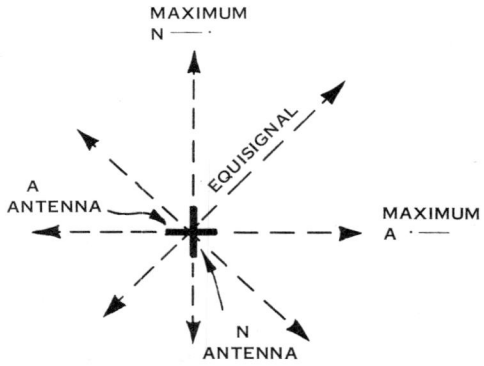

Fig. 13. Radio-range station. The equisignal is halfway between the maximum N and the maximum A.

detected by listening to the two signals in succession, defined a course leading straight to the station (or away from it). Two crossed antennas produced four courses. The Bureau of Standards erected such a station at Washington in 1921 and verified that the equisignal zone was about a mile wide 35 miles from the station. The station was then moved to McCook Field, Dayton, for further work. The main change there was to alter the time of transmitting the two Morse characters, spacing them so that the dots and dashes fitted together exactly to form a continuous monotone; the equisignal was thereafter called the beam (Figure 14). The change made it easier to detect

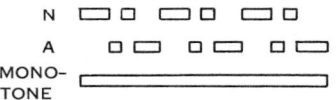

Fig. 14. The radio beam. The N and the A interlock to form the sound of flying "on the beam."

slight differences in strength of the two signals and effectively reduced the width of the equisignal zone from 8° to about 3°.[57]

Lacking both the necessary ground stations and the money to build them, the Post Office did not equip its airplanes with radio; but experience on the night mail route from Chicago to Salt Lake City, after 1924, showed that navigational aids other than light beacons were going to be necessary. The Air Commerce Act of 1926 gave the Commerce Department authority and money to install additional aids.[58]

Everyone agreed that the new aids would be radio, but which of the competing systems should be adopted?

Commerce invited the various organizations concerned to a conference in June 1926. The reasons that had caused the Europeans to adopt the ground direction-finding net were still operative. The Americans too decided to place the expensive, heavy equipment on the ground; but they carried the idea even further and adopted the range station rather than the direction-finder net. To use the net, an aircraft had to request the service, which required a transmitter. The European service had been designed to serve airliners, which were equipped with two-way radio for communications. The American system would also cater to private airplanes, which could carry a receiver more easily than two-way radio, because transmitters required more power and were more expensive.[59]

The decision having been taken, experimental range stations were built at College Park, Maryland, and Bellefonte, Pennsylvania, in 1927. Extensive flight testing followed. The first antennas were loops and subject to night effect; vertical antennas were adopted in 1932. Pilots found it difficult to hear slight changes in volume after listening to the aural signal hour after hour; a visual indicator was developed in 1930. Thereafter pilots could follow the beam by sight or sound.[60] They knew they were passing over the station when both aural and visual signals ceased, because the range radiated no energy directly above the station. Pilots used this lack of signal, called the cone of silence, to fix their positions precisely. Other points of importance along the airway, such as airports and airway intersections, were marked by low-power transmitters, called marker beacons, that radiated a narrow beam straight up. Marker-beacon signals also produced both aural and visual indications in the cockpit. When range stations were installed in the Rocky Mountains, mountain effect, the reflection of radio signals from rough terrain, was severe in some areas. Multiple beams appeared; some of them led into the sides of mountains. The beam signal alternately built and faded, as reflected energy sometimes added to, sometimes subtracted from, the direct signal. If the beam faded entirely, the pilot could mistake it for the cone of silence. This last problem was solved by placing a marker beacon at each range station. Additional low-power range stations placed in areas of severe mountain effect were of some value, because within 25 to 30 miles of a station the direct beam was strong enough to overcome reflected signals. Attempts to eliminate multiple beams by relocating the station within the same general area accomplished little. Pilots had to learn to expect false beams in mountainous terrain, when beyond 30 miles from the range. To avoid being mislead, a pilot had to compare the compass heading with the radio indication frequently; compute the airplane's position from any other possible source, such as intersections with other range courses; and generally carry mental dead reckoning and distrust radio indications that differed widely from what he anticipated.[61]

Radio ranges multiplied along the nation's airways. The first station was ready for use in 1928. By February 1931 a series of 21 stations covered the entire route from San Francisco to New York. By the summer of 1933, 82 range stations and 81 marker beacons were in operation, with 20 more ranges and 13 more beacons under construction.[62] The nation's principal airways were then defined by "radio beams along which pilots [could] fly as if the aircraft ran on rails." The American Society of Mechanical Engineers, Aeronautic Division, was told in 1929 that "the radio beacon makes navigation of the fixed routes a foolproof and automatic affair."[63]

Accident statistics did not bear out the rosy promise of 1929. The Commerce Department's accident classification system did not allow rapid identification of those caused by faulty navigation. The following categories were in use in 1932:

PERSONNEL	MATERIEL
Pilot	Power plant
Error of judgment	Structural
Poor technique	MISCELLANEOUS
Disobedience of orders	Weather
Carelessness or negligence	Darkness
Supervisory	Airport or terrain

Nevertheless, lacking contradictory evidence, whenever an airplane strikes the side of a mountain, one may assume that the pilot did not know where he was. In one bad month, from mid-December 1936 to mid-January, five U.S. airliners crashed into mountains or ridges—two near Saugus, California, within a few miles of each other. Not only inexperienced pilots got themselves lost. The previous winter the president of the Air Line Pilots Association, "complete with Boeing 247, was retrieved from a clump of willows and dusted off by some kindly yokels." In the winter of 1937-38 two more airliners were blown off course, wandered for a time, and finally landed with tanks nearly dry. In each of these accidents and near misses, the radio aids to navigation had shown themselves to be far from foolproof. But in typical American style, radio aids being inadequate, the solution commonly urged was to build more radio aids. The editor of *Aviation* urged the installation of a few ground direction-finder stations on the European model. More frequently, airline presidents called for more range stations; sometimes the recommendation was for the development of direction-finding equipment for use in the aircraft.[64]

As we have seen, this last solution had been refused on both sides of the Atlantic when the first radio aids were installed, but interest in airborne equipment continued. Military aviation needed equipment that would not tie aircraft to predetermined routes, and experimentation continued with both fixed and rotatable coils. The latter, when incorporated in an instrument for measuring the bearing of any radio transmitter, came to be called the radio compass. With the increase in off-airways flying, especially by private pilots, the commercial potential for the radio compass reinforced the military interest, and a number of models were on the commercial market by the mid-1930s.[65]

The radio compass operated on the same principle as the NC flying-boat equipment, but accomplished most of the operations automatically. The instrument became practicable for smaller aircraft only after the development of a more sensitive receiver, a static-depressing teardrop cover for the antenna on the outside of the fuselage, and a feed-back circuit that permitted automatic tracking of the station, once the antenna had been aimed at the transmitter. The Army Air Corps hired Geoffrey Kreusi, who had started work on a radio compass when employed by Western Air Express. With Air Corps money and the Wright Field, Dayton, experimental facilities, Kreusi developed the instrument to the point that the Air Corps adopted it as standard equipment in 1935. Fairchild Aerial Camera Corporation manufactured the Kreusi compass for the commercial market. Competing models by Bell Laboratories, Marconi, and, in cooperation, RCA and Sperry Gyroscope appeared in rapid succession.[66] The radio compass operated in the same frequency band as the ground direction-finders and the range stations, but the compass could sometimes be used to locate a station when atmospheric static blocked reception of the range signals. In Europe the direction-finding net was badly overloaded by 1939; at peak periods the frequencies sounded like the zoo, "especially the parrot house." Pilots were increasingly forced to get their own bearings with a radio compass.[67]

After World War II had begun, RCA announced the next generation of radio navigational equipment which would put both the range and the radio compass out of business: the omnirange. This equipment, operating in a higher frequency band, escaped most atmospheric interference. It was completely automatic in operation and more accurate than any previous directional radio equipment, but it lies beyond the limits of this book.[68]

Between the wars airline crews relied on map reading and radio

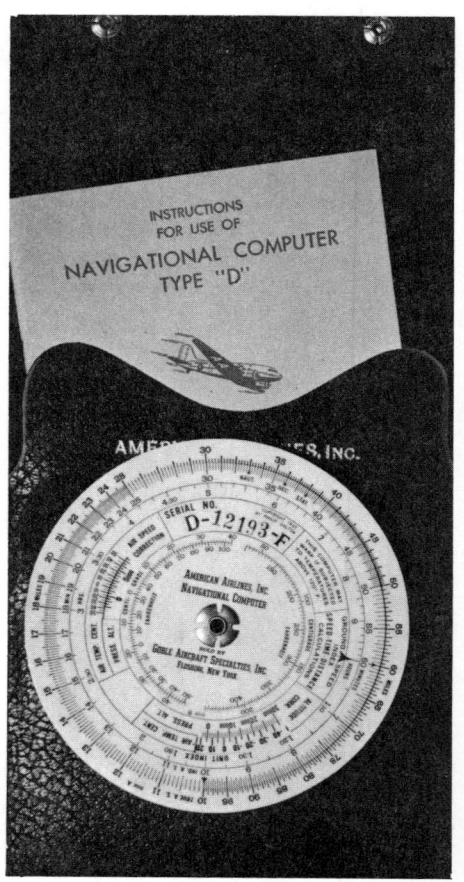

American Airlines
navigational computer,
1933.
Weems & Plath photo.

exclusively for overland navigation. Occasional critics argued that such procedures unnecessarily restricted the aerial vehicle by tying it to a railroad or a radio beam. To enjoy the maximum potential of air travel, they believed, aerial navigation had to follow "the ship rather than the iron-bound train" and cast loose from all ties to the surface. They were thinking of celestial navigation and the driftmeter as well as imagining a family of instruments not then technically feasible, gyros and accelerometers to detect motion without reference to anything outside the aircraft. The argument was sometimes stated as "to navigate or not to navigate," although that usage could scarcely be justified semantically.[69] Pilots in fact navigated if they were concerned with knowing where they were and where they were going.

And airliner pilots on domestic routes did not need to be able to choose from an infinity of routes. Rather, they flew the same routes day after day, and repeated experience on the same route produced surprising ability to recognize small landmarks and even peculiarities of individual radio stations. Such skills, almost unconscious on the part of the airline pilot, probably explain the sixth sense often attributed to experienced aviators; they extract useful information from evidence most people do not notice.[70] Map reading and radio, used carefully, served the domestic airlines well enough; but when the long-anticipated overseas airlines finally began operations in the 1930s, the critics were vindicated. Then more sophisticated means of navigation became essential.

TRANSOCEANIC AIRLINES AND CELESTIAL NAVIGATION

The crossings of the North Atlantic in 1919 and of the South Atlantic in 1922 did not inaugurate commercial transoceanic air service; aircraft and engines to carry commercial loads over such distances did not exist. Another flurry of oceanic flights in 1927 and 1928 also failed to signal immediate commercial service; by then airplanes could fly farther more safely, but almost all of the load still had to be fuel.

In 1930 the French began intermittent experimental flights across the South Atlantic, hoping to connect the overland segments they were already operating along the African and South American coasts, but their airplanes proved inadequate. The Germans made experimental South Atlantic flights with the *Graf Zeppelin* in 1931 and placed the airship in commercial service between Friedrichshafen and Rio de Janeiro in 1932. The airship flew the round trip fortnightly each summer until 1937. Deutsche Lufthansa began airplane service across the South Atlantic in February 1934; Air France followed in January 1936. Pan American, after spinning a web of routes around the Caribbean and South America, formed its Pacific Division at Alameda, on San Francisco Bay, in January 1935. By November the airline was carrying airmail to Manila, via Hawaii, Midway, Wake, and Guam. The first commercial service across the North Atlantic was flown by the *Hindenburg*. After ten round trips in the summer of 1936, the airship exploded at the end of its first crossing of the 1937 season. That same year Pan American and Imperial jointly opened the New York–Bermuda route and flew the New York–Newfoundland–Ireland route experimentally. Both Pan American and Lufthansa had airplanes capable of commercial service across the North Atlantic by 1938, but the British did not, and political difficulties prevented German-American agreement. Not until May 1939 did Pan American fly the first regularly scheduled airplane service across the North Atlantic, carrying mail from New York to Marseilles, via Bermuda, the Azores, and Lisbon. In June, Pan American flew mail on the northern route, New York–Newfoundland–Portsmouth. And in July 1939 the Clippers offered passenger service over both North Atlantic routes.[1]

On those routes that used island bases, the penalties of faulty navigation were easily imagined; a few times they were made explicit. Most aircraft lost at sea in the 1930s left no evidence for a post-mortem, but sometimes they went down while in radio contact with ground bases. On two such occasions, the aircrews were unable to find their island destinations and the radio signals were too weak for ground direction-finders to measure bearings. In 1934 Charles T. P. Ulm, an experienced oceanic aviator, left San Francisco on a trans-Pacific flight to advertise the potential of an Australia–United States airline. (Appendix B tabulates details concerning the various flights mentioned in this chapter.) He bought a twin-engine land plane and had it modified with an extra fuel tank in the fuselage and work space for a navigator–radio operator in the rear. The navigator, who had little aerial experience, could not get to the cockpit. Worse, Ulm could not see the charts and log. They planned the flight to arrive near Hawaii after sunrise, in order to have daylight for finding the islands. They did not find the islands and after four hours of fruitless radio calls to Hawaii announced they were going down, out of fuel. The next year Amelia Earhart started around the world for the fun of it. Frederick Noonan, a Pan American navigation instructor, was her navigator. Noonan had an unenviable job. Amelia Earhart obviously had an intense desire to excel in the man's world of aviation, and she had previously flown both the North Atlantic and the Eastern Pacific solo. For the small island fueling stops of the South Pacific, she concluded she had to have a navigator. Two previous to Noonan declined the offer. On the first ocean crossing of their world flight, approaching Africa, he said to turn right for Dakar; she felt a left turn was in order. She turned left, but he was right. Such lack of confidence is not the stuff crew coordination is made of. They disappeared while trying to find Howland Island, on the equator north of Samoa. Again intermittent radio contacts over six hours were too weak for the measurement of bearings.[2]

Even for those flights with mainland destinations, commercial aircraft could not afford the luxury, indulged by some private fliers, of merely aiming for the side of a continent. Passengers would not tolerate such haphazard operations. Further, time is money in commercial aviation, and overseas airliners especially could not afford to carry fuel they did not have to have. For passenger operations, airliner crews had to know their positions at all times over water so that, in the event of a forced landing, rescuers could be directed to the scene. Pan American insisted on this ability for its first overwater route, the 100 miles between Key West and Havana, in 1928.[3] Where overland airlines relied on map reading and radio exclusively, the overseas airlines concluded that their aircrews had to use all possible means of navigation. No single method was perfect, and safety was multiplied when data from different sources agreed.

Radio was not to be abandoned. It remained the only navigational aid when flying inside clouds. Pan American developed its own network of long-range direction-finding stations for use around the Caribbean and later set up a similar net to cover its Pacific route.[4] The advances in radio described in the previous chapter were applicable to overseas airlines. Because radio bearings become more accurate at short ranges, they were most useful to overseas airliners during the most critical phase of their journeys—approaching destination. Even equisignal beacons were sometimes used. For the U.S. Army flight from Oakland to Hawaii in June 1927, beacon transmitters, similar to those being placed in service on domestic routes but much more powerful, were installed at departure and at destination. The beam was of little use to the Army crew because their receiver failed, but some participants in the Dole Race, flying the same route the next month, used it to advantage.[5] But radio equipment could fail at inconvenient moments, and atmospheric static could put any radio out of business before the advent of higher frequency equipment during World War II. Radio was therefore not supplanted, but supplemented. The new techniques were weather map flying (planning a route to take advantage of the pressure systems for quicker trips) and the conversion of celestial navigation from the realm of the extraordinary to the routine. Continual improvements in the older navigational instruments also contributed directly to overseas navigation, and it is as well to take account of the changes in these instruments first.

By authoritative estimate in 1921, errors of "1 to 5 per cent were common in instruments as actually installed in the airplane," and for those instruments "used in navigation and in landing, errors of this magnitude constitute[d] a serious if not fatal defect."[6] As commercial aviation came of age, many companies catered to the instrument requirements of the aircraft manufacturers and navigators. Well-established firms, such as General Electric, Radio Corporation of America, Bausch & Lomb, and Bell Telephone's Western Electric in the U.S., and C. P. Goerz and Zeiss in Germany, formed instrument divisions. Old-line manufacturers of marine instruments, including Henry Hughes & Sons and S. Smith & Sons of London, Plath of Hamburg, and Jules Richard of Paris, entered the new field; and new companies were formed—Pioneer Instrument (which became a subsidiary of Bendix in 1931), Kollsman, Sperry Gyroscope, and Link Aviation Devices, all of New York, and Aéra of Paris. The variety of aerial navigation instruments multiplied. Not surprisingly, many putative changes advertised assiduously by the manufacturers were not in fact new. On the other hand, innumerable small changes of design, scarcely noticeable from the face of the instrument, cumulatively produced a set of instruments that were significantly better than those of 1921.

The airspeed indicator is one of the instruments in which piecemeal change through two decades resulted in a generally satisfactory instrument, while the face of the indicator, other than the higher numbers on the scale, remained much the same as the unreliable instrument of World War I. Rubber tubing, used to connect the indicator to the pressure-sampling head out on the wing, gave way to metal. Maintenance problems thereby declined. The best location for the pressure head was the subject of intensive research. An area where the airstream was relatively undisturbed was needed, and this differed from one wing to another. The pressure head itself changed. It was heated electrically, to prevent its being clogged by ice. Baffles were inserted to prevent the system being filled by water, which could happen not only in heavy rain but during every takeoff by a flying boat. Screens were added to keep out a tropical hazard, insects. On the eve of World War II a new error appeared: at high airspeeds the air is so compressed that the traditional method of comparing impact to static pressure is no longer adequate. An additional correction has to be applied by the crew. Although the airspeed meter could not be made a perfect machine, the physics of the problem was understood, and by 1939 aircraft operating manuals included graphs for correcting the indicated airspeed for the various errors. Accurate airspeed could therefore be found.[7]

At the end of World War I, the German selenium compass had been widely hailed as the shape of the future aerial compass (above, p. 71). Experience uncovered unexpected faults in the instrument. The selenium cells could not be kept electrically identical. Attempts to improve one characteristic of the instrument penalized performance in another area. Against slow but certain improvement of the traditional magnetic compass, the selenium compass was abandoned by German aviation;[8] but the problem that had led to the selenium compass—the magnetically difficult environment of the cockpit—still existed.

The next remote reading compass, developed in the United States, was the earth inductor. It was a weak electrical generator which used the earth's field instead of the usual magnet to supply the magnetic

The earth inductor compass, not installed. McCook Field, September 1924. *USAF photo.*

lines of force. The principle had long been used in magnetic surveying instruments. By 1924 the Bureau of Standards and the Army Air Service's Engineering Division, McCook Field, Dayton, had brought the instrument to a practical form for use in aircraft. An experimental model, manufactured by Pioneer Instrument Company, was tested on the U.S. Army round-the-world flight that year. A small windmill mounted on the top or side of the fuselage turned the generating coil. The dashboard instruments resembled those of the selenium compass. An indicator showed whether the aircraft was on course or off to the right or left, and a course-setting control was also provided. The pilot

Instrument board of *Spirit of St. Louis*.
USAF photo.

reports were favorable, and Pioneer had an improved model in quantity production in time for the long-distance flights of 1927 and 1928.⁹

The Pioneer earth inductor was installed on most of the airplanes making those flights. Lindbergh, enjoying his triumph, lavishly praised the instrument: "Laymen have made a great deal of the fact that I sailed without a navigator and without the ordinary stock of navigation instruments, but my real director was my earth inductor compass. . . . which guided me so faithfully that I hit the Irish coast only three miles from the [intended landfall] . . . . This compass behaved so admirably that I am ashamed to hear any one talk about my luck. . . . This inductor compass was so accurate that I really needed no other guide." Some of the other aviators who used the inductor had a different story to tell. Clarence D. Chamberlin intended to rely on the instrument on his flight from New York to Germany in June 1927; it failed over Cape Cod.¹⁰ Richard E. Byrd's *America* had the same instrument on the New York–Paris attempt the same month. The needle of the indicator began sticking (and the radio failed) after crossing the French coast. The weather closed down over Paris and the aircraft was successfully ditched near Caen, just off what would later be famous as Omaha Beach.[11] The earth inductor served George Wilkins and Carl Ben Eilson well on their flight from Point Barrow to Spitsbergen in April 1928. Kingsford-Smith and Ulm had nothing but praise for the inductor after their remarkable flight from San Francisco to Australia, via Hawaii and Suva, in May and June 1928, although the instrument failed them on the last leg. The crew took responsibility: they had failed to oil it properly.[12]

The record of the earth inductor was therefore spotty. Pioneer of course exploited Lindbergh's testimonial, but the instrument had not in fact given him exceptional service. The first instrument installed in the *Spirit of St. Louis* failed on a test flight at the San Diego factory. Then, in mid-Atlantic, as an older Lindbergh would recall, both the inductor and the secondary magnetic compass had swung widely for much of the night. Lindbergh had steered by the stars for long periods, when they were visible, and sometimes found the secondary compass more reliable than the inductor when the stars were hidden.[13]

What was wrong with the inductor compass? The problem of holding the generating element horizontal was never solved. A pendulum kept it more or less horizontal, but enough has been said about the difficulties of defining the vertical in flight. The inductor not only required a universally mounted sensing element; power had to be applied to that element without introducing an error. The generator brushes, used to pick off the directional signals, had to be several degrees wider than the desired accuracy of the indicator, and the brushes wore unevenly. German firms, having abandoned the selenium compass, experimented with its successor, the inductor, for a time, but soon abandoned the inductor as well.[14] After 1932 nothing more was heard of the inductor.

The next advance in aircraft compasses depended on the gyro. The gyro itself cannot differentiate between one direction and another, but it can retain the same direction for a period of time, in spite of aircraft motions. For use as a compass, a magnetic element combined with a gyro for stability was an obvious step. Such an instrument, built by Sperry, appeared on British warships in World War I, but it was far too heavy for use in the air. By 1920 Sperry had a gyro compass small enough for use in aircraft, but its U.S. Air Service test was unsatisfactory. The work continued on both sides of the Atlantic, and the *Graf Zeppelin* used such a compass with good results on its world flight of 1929. The instrument was still too heavy for airplanes, but an alternative solution was adopted: use two separate instruments, a directional gyro to steer by and a magnetic compass to reset the gyro periodically. Sperry offered such a gyro in the early 1930s. Wiley

Post and Harold Gatty used the combination on their world flight in 1931, as did Amelia Earhart on her Atlantic flight the following year. The Lindberghs then tested the directional gyro on their extensive survey flights for Pan American in 1933 and found that steering by it improved the accuracy of celestial observations. The instrument became standard equipment on the Clippers.[15]

The two instruments were finally combined successfully in 1937, when the U.S. Air Corps found the Sperry Gyro-Mag sufficiently advanced to place a service test order. Previously gyros had been mounted on steel ball bearings, but the whirling steel bearings could not be used near a magnetic compass. A major problem was therefore a new bearing; Sperry worked almost ten years to perfect air bearings, supporting the gyro on compressed air. A magnet on the gyro motor housing controlled two small air jets that kept the compass card aligned with the magnetic meridian. The gyro held its horizontal position during turns, so that turning errors were finally eliminated from the magnetic compass. An early model of the instrument was in use experimentally on Pan American's Clippers by 1938. Sperry hoped to offer a commercial model by 1941, but by that time instrument production was diverted for military purposes.[16]

Back in the 1920s, several flights had been made in the Arctic and Antarctic. Near the earth's magnetic poles, the magnetic lines of force become vertical and there is nothing for a magnetic compass to measure. Although improved magnetic compasses were found to operate satisfactorily much closer to the magnetic poles than had earlier been generally assumed, a new instrument for directional reference was needed for polar flights. The gyro compass would have been welcome, but it was still too bulky. Because the exploratory flights were conducted during the long polar day, an instrument that would determine direction from the bearing of the sun was an obvious solution. In 1917 an American patent had been issued for a transparent sundial mounted in the wing above the pilot, but it does not seem to have been put into production.[17] The idea was sound, but further development was needed.

Two practical sun compasses were designed in the mid-1920s. Albert H. Bumstead, the Chief Cartographer of the National Geographic Society, devised one for Commander Byrd's use in northwestern Greenland and Ellesmere Island in 1925. The instrument consisted of a clock running on solar time, with a single hand which made one revolution every 24 hours. The clock's mount could be tilted to align it with the horizontal at the pole, which required only an approximate knowledge of the aircraft's latitude. The tip of the clock hand carried a vertical pin which, in the sunlight, threw a shadow onto a compass rose. The rose itself could be rotated, to place the desired heading under the shadow, after which the pilot steered so as to hold it there. A similar instrument, manufactured by the Goerz firm of Germany, was first used by Roald Amundsen and Lincoln Ellsworth in the Arctic in 1925. The Goerz sun compass used a clock to aim a periscope toward the sun. An optical system then transmitted the sun's image to an opaque disk that carried cross hairs, and the pilot steered to keep the image beneath the vertical hair. Both instruments were theoretically sound, relatively simple to use, and subject to an obvious shortcoming. As Byrd wired the Navy Department in 1925: "Sun compass very good when sun is visible."[18] While the exploratory flights were undertaken only during the long polar summers, weather forecasting was almost nonexistent. Furthermore, locating the instruments so that the aircraft structure would not block the sunlight was not easy. On their flight from Spitsbergen to the pole and back in 1926, Byrd and Floyd Bennett used two Bumstead compasses. One was mounted permanently on a trapdoor in the top of the navigator's cabin. When the wing shadowed this area, Byrd used a second, portable instrument through one of the windows. For crossing the Arctic Ocean in 1926 in the airship *Norge,* Amundsen and Ellsworth provided mounts for their Goerz sun compass on outriggers on opposite sides of the control gondola. "It was the navigator's most unpleasant job to shift the compass from one side to the other, as this must be done with half his body outside the gondola in an 80-kilometer wind. . . . To manipulate small screws in such a wind and at such a temperature is no agreeable task." Nevertheless, throughout the 1930s one or the other of the sun compasses was essential on polar flights. The Russian crew of the ANT-25 apparently carried one of each on their Moscow–Washington State flight in 1937.[19] During World War II, the astro compass, a derivative of the sun compasses, would permit navigators to determine aircraft heading from any celestial body whose position could be computed from an almanac. The astro compass could also be used with a partially obscured sun.

With improved magnetic compasses, the sun compass, and the various applications of the gyro, navigators could determine direction with increasing precision. But knowing the direction in which the aircraft was headed was of course not enough; the effect of the wind had to be measured. The driftmeter, abandoned by commercial and private overland fliers, remained of importance for overwater navigation. The Germans merely adapted their bombsights to the new "eminently peaceful purpose." For Lufthansa's Atlantic aircraft, Zeiss produced a hybrid instrument, incorporating a driftmeter and DR computer (similar to the Wimperis wartime bombsight, p. 79) with a compass and altimeter, so that all the data necessary to find the wind was handily assembled; but except for the fetching name Quo Vadis, it was no particular advance.[20] The French, however, attacked one of the principal defects in the driftmeter.

Unless the air was perfectly smooth, the navigator attempting to measure either drift or groundspeed by tracking an object with the driftmeter was hampered by every slight motion of the aircraft. The value he measured unfortunately depended on both the wind and the stability of the airplane. Even if he succeeded in measuring the drift or groundspeed correctly, it was valid only for that moment. The navigator needed to know the average drift and groundspeed over a short period of time. A French naval officer, Yves le Prieur, attached the parallel arms of a pantograph, a pencil, and a pad of paper to the driftmeter. Then, instead of trying to align an etched line in the optical system with the path of an object on the ground, the navigator followed the object across the field of view with the pantograph.

The Gatty drift indicator, 1933.
USAF photo.

Actually he followed several objects in succession; each resulted in a line on the paper, usually a wavy line, but the general trend of all the lines, compared to a compass rose printed on the paper, gave an accurate measurement of the mean drift. The instrument was used on the Paris–Brussels–Amsterdam airway in the early 1920s and the French were still installing it in their large flying boats for the South Atlantic run in the late 1930s. The U.S. Air Service pronounced it as accurate as any driftmeter known in 1925. It was well adapted for finding a wind from drifts on successive headings (Figure 10), which avoided the error introduced by inaccurate altitude.[21]

Le Prieur's Navigraph was a large instrument. In the United States efforts were directed toward the design of a driftmeter suitable for a small airplane, perhaps one that a pilot could use when flying solo. The Bureau of Standards produced a driftmeter that incorporated a rotating hexagonal prism. The operator adjusted the speed of rotation of the prism until the apparent motion of objects on the ground was neutralized. Then, after inserting the aircraft altitude on one dial, he could read the groundspeed directly on another. The U.S. Navy experimented with a driftmeter that would indicate the drift on a dashboard repeater. Neither instrument seems to have been satisfactory.[22]

Harold Gatty,* one of the few professional aerial navigators of the 1930s, approached the problem from a different way. For his world flight with Wiley Post in 1931, he designed a periscopic driftmeter that used a constant speed film, transparent except for evenly spaced parallel lines at right angles to the motion of the film. By adjusting the height of the eyepiece above the film, the navigator synchronized the motion of the film with objects on the ground. The height of the eyepiece above the film could then be converted to groundspeed, if the altitude was known. The navigator did not attempt to follow a particular spot on the surface, but learned to match the lines on the film to the "flow" of the total landscape. This feature made for greater accuracy over water, where sighting points were few and indistinct. But the instrument still suffered from the need to know altitude; Post and Gatty more than once descended almost to the surface, set their pressure altimeters, and climbed back before measuring the groundspeed.[23]

---

* Born in Tasmania in 1903, Gatty was a self-taught marine and aerial navigator. He opened a navigation school for pilots in San Diego in 1928, was associated with the Weems System of Navigation from 1928, and was appointed technical adviser to the U.S. Army Air Corps in 1932. He was an honorary Group Captain, RAAF, in World War II and died in Fiji in 1957.

Gatty's driftmeter did not supersede the traditional instruments that used wires or etched lines in the optics. Pan American used a variety of models. For taking sights over the ocean, both Pan American and the RAF found a better artificial landmark than the floating smoke flares of World War I: bronze or aluminum powder dropped into the sea spread quickly into a wide, bright spot that could be seen for many miles. The navigator dropped the bomb of powdered metal and then tracked the spot aft, to measure both drift and groundspeed. At night, flares still had to be used.[24]

The ubiquitous gyro was added to the driftmeter in the late 1930s. By mounting the reticle of the optics on a gyro, the apparent wandering of ground objects caused by the aircraft's pitching, rolling, and yawing was eliminated. The spatially stabilized driftmeter improved DR accuracy noticeably. The U.S. Army Air Corps was installing gyro driftmeters in the B-17 by 1938.[25] The remaining item needed for precise groundspeed measurement with the driftmeter—the aircraft's altitude above the terrain—was available shortly thereafter.

The altitude needed could not come from the pressure altimeter, although work on this instrument continued between the wars, and a number of minor improvements together produced a markedly better instrument.[26] A completely new device was needed, something that would measure the space between the aircraft and the surface of the earth, not the pressure difference.

The 1st Air Conference of the British Empire, meeting in London in 1920, was told that some aircraft accidents could be avoided if there were an instrument that could "record the actual height of a machine above the actual ground level. . . . Whether such an instrument is within the realms of possibility," the speaker concluded, "I do not know, but if it could be produced there will be general agreement that machines in future should be fitted with it." Such an instrument was produced, but it took two decades. The first attempt to produce an absolute altimeter (to distinguish it from the pressure altimeter) was announced at the same conference by a later speaker. Two large plates attached to the bottom of an aircraft wing were connected to form a large capacitor. When the aircraft came near a third conducting surface, the sea or the earth, the capacitance of the circuit changed. The device could not be developed, because the electrical changes were slight and an unexpected variable entered the equation: wet earth caused a greater change than did dry.[27]

The Germans next entered the field with acoustical altimeters, which were made to work reasonably well on airships at moderate heights. The Germans were interested in the instrument because of its potential for weather-map or pressure-system navigation, which was pioneered by Hugo Eckener and his Zeppelins. The airship had such a low airspeed that head winds had to be avoided wherever possible, even to choosing a much longer route. This led those dealing with airships, a field in which the Germans maintained their wartime lead, to be concerned with the implications of atmospheric pressure systems for long-range flight before others thought much about it. In the Northern Hemisphere the winds blow counterclockwise around areas of low pressure, clockwise around highs. A flight to one side of a given pressure system will therefore encounter a head wind, to the other side a tail wind. When the British airship *R 34* made its round trip of the North Atlantic in 1919, a meteorologist was on board to interpret the weather data received by radio and to make on-the-spot observations. When the centers of the pressure systems could be located, the route was altered to take advantage of favorable winds.[28] The pressure systems could also be located by taking pressure measurements near the surface periodically. Airships could either descend for these measurements or a recording aneroid could be lowered on a line. If, however, the crew could measure the absolute altitude and pressure altitude simultaneously, at flight altitude, the difference between the two would also indicate surface pressure. The Germans were therefore interested in developing an absolute altimeter for use on their civil airships.

When Hugo Eckener delivered the *LZ 126* to the United States in 1924 (it was the American share in reparations), he had two experimental absolute altimeters on the airship. One was an optical device, in which a spotlight beam was bounced off the water and the angle measured with a sextant. Unsuitable for cloudy weather, the optical altimeter was of limited value. The second was an acoustical altimeter, one of the first of a long line. A cartridge was detonated just below the airship and a sound receiver started a clockwork mechanism that drove a pen over a graph. Receipt of the echo stopped the timer. The elapsed time, read on the graph, was then converted to distance—height—with allowance for the round trip. The instrument suffered from two faults: the speed of soundwaves in the air is not constant, and any sound would stop the timer.[29] In subsequent developments, the instrument was freed from the second fault, by designing a sound receiver that filtered out all but the desired frequencies. Because of the need for an absolute altimeter, both in navigation and in landing through fog, the sonic altimeter received widespread attention, but the obvious difficulty of using any sonic device in an airplane was never overcome.[30] Fortunately, by the mid-1930s a radio altimeter promised to fill the same need.

Instrument board of Douglas BT-2B fitted for blind flying at Randolph Field, March 1932. Note compass correction card, sonic altimeter, Sperry gyro horizon, and directional gyro.
USAF photo.

Radio waves could be reflected from the surface of the earth, but the problem was to measure the time interval, which was extremely short. The first attempts used very low frequencies with correspondingly long waves, with equipment to measure the maximum and minimum strength of the reflected signal. An 8,000-foot wave gives a maximum signal at 2,000 and 6,000 feet, minimum signals at 4,000 and 8,000 feet. As an altimeter, an instrument operating on this frequency would give the same indication every 4,000 feet, but the pressure altimeter could be used for an approximate measurement before referring to the radio altimeter for a precise one. The trouble was that very low frequency waves were largely absorbed by the earth; few returned to the aircraft. At higher frequencies, the shorter waves reflected well, but the intervals between maximum and minimum signals were much closer together, too close to be differentiated by reference to a pressure altimeter. The attempt to measure maxima and minima was therefore abandoned, and attention turned to the transmission of a wave whose frequency was continually changing at a known rate. A sample of the returned energy and a sample of the wave currently being transmitted were fed into a frequency meter, which detected the difference between the two. This meter was calibrated to read altitude in feet. Radio altimeters operating on this principle appeared in several countries in 1938 and 1939 and went through the usual developmental steps, as engineers sought greater reliability with less weight.[31] The simpler system of using short bursts of radio energy with an oscilloscope for measuring the interval between transmission and reception of the echo would appear during World War II. The radar circuits required to form the short pulses existed in 1939, but they were carefully guarded military secrets and not available to those engineers and airlines interested in absolute altimeters.

Even without a practical absolute altimeter, the transoceanic airlines planned their long flights to take advantage of the pressure systems, using the best weather forecasts available. Byrd planned his Atlantic crossing in close coordination with the U.S. Weather Bureau office in New York, but in 1927 very little was known of weather over the Atlantic. Some ships took weather observations and, upon reaching port, mailed them to the bureau. The forecasters used this information to evaluate their own forecasts of the previous weeks. Byrd was in clouds for most of his crossing, therefore unable to do much toward changing his route to take advantage of favorable winds. His log read hour after hour, "It is impossible to navigate." Crossing the Atlantic the next year in the *Graf Zeppelin,* Eckener found how unreliable weather reports, even comparatively recent, could be. After receiving a weather report from Newfoundland by radio one morning, the wind unexpectedly increased in velocity and shifted direction that afternoon. As a result, while flying in clouds he was blown 170 miles off course in less than three hours, a significant navigational error by any standards. Continuing that flight around the world, Eckener discovered that it was impossible to obtain weather information over most of the globe. When he could do so, he planned his flight to take maximum advantage of favorable winds, even pushing into known storm areas for the sake of high tail winds. Pan American always planned its Pacific flights on the weather map and in the first three years flew the great circle route between San Francisco and Honolulu only four times; on all other occasions, a longer distance could be flown more quickly—if the forecast was correct. And by 1938 enough meteorological information was available for the North and South Atlantic that Donald C. T. Bennett, flying from Scotland to South Africa, planned a single-heading flight. That is, he averaged the forecast wind effects for the entire route, found a net drift, and allowed only for that value. The wind blew the aircraft off course to one side and then the other; but to the extent that the forecast was accurate, this procedure was the most economical. The airplane did

Lt. Commander Weems using a marine sextant with bubble attachment. *Weems & Plath photo.*

not expend energy fighting the wind but rather allowed the wind to blow the airplane back and forth across the desired route, all the while converting its own energy into forward motion.[32] An absolute altimeter would have allowed Bennett to correct his planned heading on the basis of actual conditions encountered in flight; but, as explained above, such equipment was just appearing in practical form at that time.

All the instruments described above—long-range radio, airspeed indicator, magnetic and gyro compass, driftmeter, pressure and absolute altimeters—were necessary for transoceanic navigation. But the single instrument most closely associated with long, overwater, commercial flights between the wars was the sextant. With all its shortcomings, and they were severe, the bubble sextant still offered the navigator a means for finding his position in mid-ocean, without relying on any ground agency. High clouds sometimes prohibited observations, but on a long flight, chances were good that breaks would appear in the clouds sooner or later. During the day, the sun was often visible through a thin overcast. And a celestial fix, if not too inaccurate, wiped out all the accumulated error in the dead reckoning. These reasons, valid enough, do not seem quite to explain the attachment navigators have for celestial navigation. Perhaps the navigator sees in the sextant his *raison d'être,* for of all his methods, celestial navigation is the most incomprehensible to a pilot *qua* pilot. There is something else, however; and Patrick Gordon Taylor, a self-taught Australian navigator, expressed it in writing of his first encounters with celestial navigation: "And so I began to *see* the heavens . . . as a vast and intimate revelation which I was beginning to understand, and with which I already felt the first touches of an immensely satisfying contact. It was the dawn of a security in the air coming back to me from the infinity of space, instead of from a check back to objects or communications on the Earth."[33] Whatever the reason, navigators on overseas demonstration flights used celestial navigation, stressed its importance, and encouraged the development of newer instruments and techniques; and all the transocean airlines required their navigators to use celestial positioning. The instrument manufacturers were slower to fill the need for improved sextants than for other devices, because the unit cost was high, the market thin, and profit potential therefore slight.

Philip Van Horn Weems,* founder and until recently principal owner of Weems System of Navigation, Annapolis, has been associated more closely than any other individual with the promotion of celestial navigation for airplanes. He has addressed aeronautical meetings, contributed to the aeronautical press, written textbooks, and kept up correspondence with inventors and manufacturers both here and in Great Britain. While his firm dealt in all navigational instruments,

---

* A graduate of the U.S. Naval Academy in 1912, Weems was navigator on USS *O'Brien* during the NC trans-Atlantic flights of 1919 and on USS *Rochester* from 1921 to 1924. He taught navigation at the USNA from 1924 to 1927. While stationed on the West Coast, 1928 to 1931, he pursued problems of naval aircraft navigation on and off duty. He attended Postgraduate School at Annapolis and was associated with the Hydrographic Office until retirement in 1933. He has been occupied with full-time commercial exploitation of the needs of aerial and marine navigators ever since. Weems has been periodically recalled, not only for sea duty in World War II but to supervise navigational research projects.

celestial navigation has occupied a special place in his interests. As late as 1936 he was still trying to convince overland airlines to use celestial navigation. But in spite of his best efforts, he had to "count more on other branches of air navigation, and on celestial navigation for the mariner" as sources of income until the approach of World War II.[34] Weems was unable to sell celestial positioning to overland fliers, but the overwater navigators needed no selling. What they needed was a better sextant, along with improved reduction methods—the same difficulties that had plagued aerial celestial navigation since the days of the balloon.

The aerial sextants available in the mid-1930s had changed little from those of the early 1920s—the R.A.E. models in Britain, one by the Bureau of Standards in the United States, and the Plath-Coutinho in Germany (above, pp. 107–13). The bubble continued to give trouble. The sextant is exposed to extreme changes of pressure and temperature, and this was even more true before the days of pressurized cabins. Manufacturers took two different approaches in designing the bubble. One solution was to fill the bubble cell not quite full of liquid, leaving enough air for the bubble. The alternative was to fill the cell completely, seal it, and then increase its volume by expanding a flexible diaphragm. Liquid being unable to expand, part of it had to change to a vapor bubble. The former method produced a fixed bubble, the latter an adjustable one. The adjustable was the more susceptible to changes in temperature and pressure, but neither was satisfactory. The bubble changed size during an observation, which introduced error, or it disappeared altogether. The sealed units ruptured too often. On the first leg of the *Southern Cross*'s trans-Pacific flight in 1928, the bubble failed; it was repaired in Hawaii. Lindbergh had to puncture the bubble element of his sextant with a pen knife on his 1933 Atlantic survey flights, using the natural horizon thereafter. Navigators of the U.S. Air Corps and Navy condemned their respective service models in 1934 and 1935. The Russian flight over the North Pole to Seattle in 1937 had similar difficulties: the navigator found he could heat the sextant against the engine radiator and take a quick observation before the liquid cooled and the bubble disappeared again. An official German publication of the same year complained of the difficulty of keeping a bubble from changing size during an observation. Weems urged sextant manufacturers to attend to this shortcoming. In 1934 he wrote Pioneer: "I make one urgent suggestion: that you make the bubble element so strong that novices can't damage it with a stillson wrench." The problem was not solved before the war; in 1940 Link Aviation Devices produced a sextant with an easily removable bubble cell and supplied spares, so that the unit could be replaced in flight.[35]

Most celestial navigators concluded that accuracy with the bubble sextant depended on two things: averaging a large number of observations (the practice of the pioneer balloonists—above, p. 31) and having a pilot who would fly as straight a path as possible, eschewing power changes, during the period of observation. Several navigators attributed their success to careful piloting. Advances in compasses helped pilots hold a smoother course. Lindbergh found in 1933 that steering by the directional gyro improved the accuracy of celestial sights. The Sperry autopilot, installed on the later Clippers, flew the aircraft smoother than a human pilot, and celestial observation benefited again. But as the balloonists had discovered, people moving around in an aircraft change its center of gravity and disrupt celestial observations. Passengers on the Clippers milled about as they saw fit. The Boeing 314 Clipper therefore was equipped with a warning light inside the cabin, by which the navigator requested the passengers to sit quietly while he took celestial sights.[36] Sextant errors due to aircraft accelerations, other than those caused by uncooperative pilot and passengers, seemed to be random; but many of them were related to oscillatory motions of the aircraft, especially the slow rise and fall of the wingtips. If a series of observations extended through the period of aircraft oscillation, the average of the observations, applied to the mid-time of the series, should give substantially better results than any single sight alone.

Navigators debated the number and spacing of individual observations. In 1929 the *Wissenschaftliche Gesellschaft für Luftfahrt*, Germany's most knowledgeable organization in aeronautics, recommended an arithmetical average of a series of six observations. Maurice Bellonte, the navigator on the first direct flight from Paris to New York in 1930, evaluated a series of observations by plotting them against time. Before beginning, he computed the expected slope of the graph from the known motion of the body.[37] In 1935 Weems studied the problem at length, using data gathered on a flight from Alabama to Virginia. He took 11 groups of ten observations each of the sun, noting the position of the airplane by map reading between each series. While one sight was 122 miles in error, the average error for the total 110 sights was 3 miles. He found that any 50 could be relied upon to produce a position line accurate within 5 miles, "that is to say, less than half the admitted error for the position of the British fleet at the Battle of Jutland." Shots that were far out of line with the others should not be discarded; eliminating the two observations that differed most from the average of the remaining eight of each group increased the error by 1.3 miles. The air had been "fairly smooth, except for occasional short periods of considerable turbulence"; the average time for each group of ten shots, seven minutes. Obviously a navigator could not afford to measure the height of a single body 50 times, using 35 minutes for observation alone. But that same year, 1935, Kollsman Instrument Company, Brooklyn, had a mechanical averager for sextants under development.[38]

By 1936 three different averagers were under development: by Kollsman, by Henry Hughes of London, and by Lieutenant Thomas Thurlow,* U.S. Army Air Corps, Engineering Division, Dayton. The principles employed by all three were similar enough to produce patent squabbles. To use the averager, the navigator first chose an altitude slightly below that of the body he was about to observe and set a mechanical stop in the sextant to mark the reference altitude. He then took three separate height observations. When the body and bubble were aligned, he pushed a button, causing the averager to add 1/6 of the height in excess of the reference altitude. After the third reading, the navigator noted the time and then took three more observations. The value from the averager added to the reference altitude was the average height for the mid-time of the six observations.[39] The engineering problems were more perplexing than anticipated. The factory manager at Kollsman had written Weems in October 1935 that the averager was "well on the way of being 'a fait accompli,'" but in May 1938 Kollsman was still "working on an averaging device for a sextant." By that time, Hughes had delivered a few Husun Mark XII averaging sextants to Imperial Airways for test and was offering the instrument commercially. Fairchild Aerial Camera had produced experimental models of the Thurlow instrument, but was not yet in commercial production.[40]

In view of the difficulties with the averager, a simpler means to the same end had obvious appeal, and again a number of persons seem to have hit on similar ideas at about the same time. The Favé-Lepetit recording sextant, manufactured by A. Lepetit of Paris, appeared in 1938. In lieu of an averager, the sextant carried a circular, celluloid disk and pencil, arranged so that, when the body and bubble were aligned, pushing a button made a mark on the disk. The navigator first noted the time, then took a series of sights, and again noted the time. He then inspected the pattern of pencil marks, visually esti-

---

* Thurlow was born in California in 1905 and attended Stanford University. He entered the U.S. Army in 1928 and was rated as a pilot in 1930. He was navigator for the Howard Hughes World Flight, 1938, and was occupied with various research and development assignments at the Engineering Division, USAAC, from 1936 until his death in the crash of an A-29 at Love Field, Dallas, in 1944.

Colonel Thomas L. Thurlow, USAAF, taking a celestial observation through an astrodome in World War II. From this photograph was designed the plaque for the Thomas L. Thurlow Award, sponsored by the late Sherman Fairchild and awarded annually by The Institute of Navigation to recognize outstanding contributions to navigation.
USAAF photograph, courtesy of Mrs. Thomas L. Thurlow.

mated the center of the group, and used this value for the average height, along with the average of the start and stop times. Two months after this instrument was illustrated in the French aviation press, Edwin A. Link, of Link Aviation Devices, wrote Weems: "It would seem to me that a sextant designed so that it would print the altitude readings and time readings simultaneously on a roll of paper which could be operated merely by touching a button, would be a vast help. . . . I have made some sketches . . . and . . . have even had our patent attorney make a search . . . . So far we have not found anything. Do you know of anything that has been done?" In December Thurlow wrote Weems: "I have always regarded the averager as a 'stop gap.' I am now modifying an instrument with a simple attachment which I have reason to believe will be superior to the averager . . . . By the end of a couple of months the device will have proved or disproved itself."[41] In February 1939 Weems made some sketches of his ideas for such a device—it would be called a

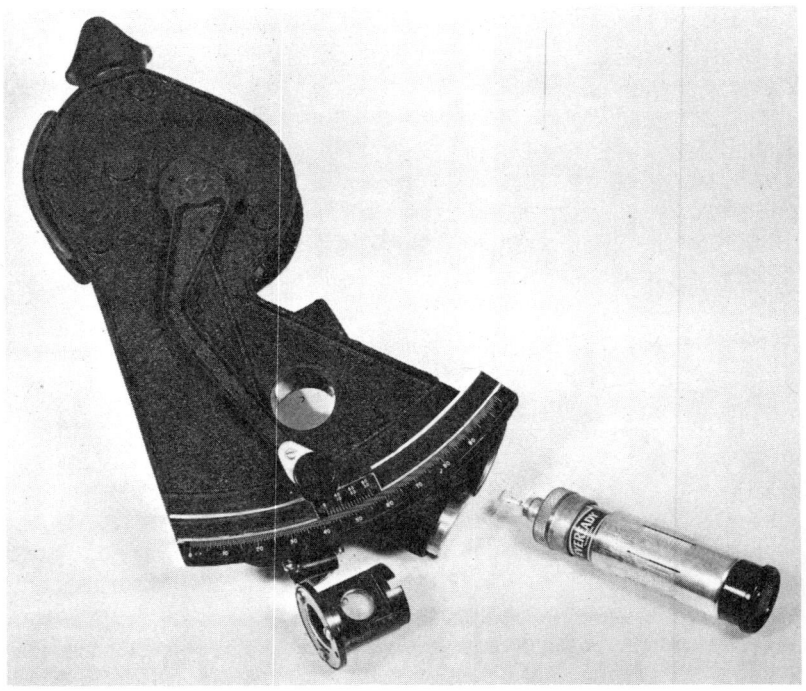

Link bubble sextant.
Link Aviation Devices photo, courtesy of Capt. Weems.

median recorder to differentiate it from the mechanical averager—and urged Link to incorporate it on the sextant Link had under development. Link replied that David Waghorn, an RAF officer, had suggested a similar device the previous year. "It looks like a good idea and I think could be worked out in a very practical manner."[42]

Link, an eminently practical man, proceeded to produce his sextant with a median attachment and was delivering it in quantity before Pearl Harbor. Pioneer produced median sextants for the U.S. Navy in the same period. Fairchild continued work on the averager as well as a median model.[43]

The Allies entered the air war with median and averaging sextants of American manufacture and the early Hughes averager. The British stayed with the averager and during the war Hughes carried the instrument two steps further. Because the navigator often chose the wrong time to take the individual measurements, being unable to differentiate between periods of zero acceleration and constant acceleration, the Hughes engineers modified the averager to run for a period of two minutes. Every two seconds the averager of the Mark IXA measured the altitude shown on the sextant at that moment and carried 1/60 of the value to a counter. After two minutes, during which the navigator tracked the body constantly, 1/60 of each of 60 readings had been automatically added on the counter. A shutter fell across the field of view, telling the navigator the two minutes had expired; he read the average height on the counter and applied it to the mid-time of his observation, which was one minute after he began, or a minute before he ended. This instrument suffered excessive wear from the frequent starting and stopping of the mechanism. The next step, which had been predicted by Thurlow in February 1941, was a continually running integrator. Captain Paul Gray of Pan American invented an integrator that he attached to his bubble sextant. Before the war ended, Hughes engineers had successfully built an integrator into their bubble sextant. Since then the hand-held bubble sextant has not changed substantially.[44]

While the averagers were being pushed to a practical stage, a new error afflicting bubble sextants was discovered. In late 1939 Thurlow heard "indirectly," apparently through a third party, that John Q. Stewart of the Princeton Observatory had suggested the possibility of Coriolis's acceleration affecting celestial observations. Coriolis, a 19th-century French engineer, had shown that a body moving over the rotating globe appears to be deflected to the right in the Northern Hemisphere and that the amount of deflection is proportional to the sine of the latitude and to the speed of the body. Artillery officers had long applied the theory in long-range fire, and weather forecasters applied it to explain wind flow. Stewart suggested that the theory might also apply to the liquid in the sextant, and therefore to the bubble. Thurlow grasped the implications and "immediately got busy, resurrected [his mechanics textbook], investigated the matter thoroughly and derived the equations for the acceleration," after which he arranged for test flights to verify them.[45] For an aircraft flying at 40° north latitude, 200 knots groundspeed, the maximum error is 3 miles. In telling Weems, Thurlow poked a little fun at those navigators who had claimed an ability to "consistently hit the point time after time, right on the nose, as I understand many can do (but I can't)," without correcting for the newly discovered error. Weems's immediate response to the news well illustrates the man's role as self-appointed popularizer of aerial celestial navigation. Weems urged Thurlow to write an article for the *U.S. Naval Institute Proceedings,* wrote the Bureau of Standards requesting verification of the theory, forwarded the equations to one of his regular correspondents in the RAF (asking that he "not give the source of this material, since

Using the sextant through a hatch in the Fokker Tri-Motor, 1927. *USAF photo.*

there might be kick-back here in passing out what is really data from official sources,") and informed Link, Hughes, and Pioneer.[46]

The averagers were improving the accuracy of sextant observations; the application of Coriolis's theory removed a previously unexpected error. But another major source of error remained: the physical problem of using a sextant in an airplane. The post-World War I navigators had worked in open cockpits. Their view of the heavens was excellent, though they had difficulty holding the sextant in the airstream. As speeds and altitudes increased, cockpits were enclosed, and the navigator lost much of his view. Hatches were then provided, allowing the navigator to stick his head and sextant outside, but this was a less than perfect expedient. The navigator not only faced the elements, manipulating the sextant with numbed fingers and watery eyes; he also had to live with the other members of the crew. A radio operator on a New York–Bermuda flight in 1930 good-naturedly complained: "About this time [the navigator Lewis A.] Yancey wanted to take a sight, which was a most efficacious way of shutting down the radio. In taking sights, Yancey opened the top of the plane, admitting a hundred-mile-an-hour blast of cold wind that wreaked havoc with the papers in the radio cabin, to say nothing of the good right hand." Radio operators think as much of their code-sending hands as do concert pianists. Conditions had not changed in the RAF by 1938: "Anybody who opens a chink of window at 15000 ft on a cold night is pretty unpopular with the rest of the crew.... Some of our present day gun turrets would make very good positions for sextant work, but for two things—they are full of guns and in war time would also be full of gunner—so they won't do. The navigator must have a place of his own."[47]

Aircraft designers first tried to satisfy the navigator's needs with optically perfect, flat windows, but the view was too restricted. Two possibilities offered: to give the navigator "a place of his own" in a special turret or dome, large enough for head, shoulders, and sextant to project above the top of the fuselage; or to design a periscopic sextant, only the tip of which had to extend into the airstream, and let the navigator work inside the aircraft. The Germans tried the latter approach.

Using a design by a retired naval officer, Wilhelm Opitz, the Plath Company produced a periscopic sextant that was tested in 1929. Strong vibrations made the bubble unusable; after adding a gimbal mount and rubber shock absorbers, a second model was tested in 1930. The optics followed the pattern established by Gago Coutinho (above, p. 111), with two bubbles in bent-glass tubes, the tubes being perpendicular to each other. One bubble was used to level the instrument fore and aft, the other laterally. To align a single body with two bubbles was not easy; the image of the body was therefore reflected through a prism to produce a double image in the field of view. The navigator then had two suns to align with two bubbles. To complicate matters further, Plath also reflected a portion of a magnetic compass rose into the field of view, so that the navigator could see the current compass heading. And finally a portion of an azimuth circle that was fixed to the ceiling of the aircraft was also visible in the field of view. The last two items were useful innovations. With them, a navigator could measure the true heading of the aircraft with reference to a celestial body, note the compass heading, and compute compass deviation quickly while in the air. Although the double bubble had been used in Plath instruments before, this feature may have contributed to the relatively inaccurate height measurements when the sextant was tested by the *Deutsche Versuchsanstalt für Luftfahrt*. The Plath Company had found the average error of 40 observations to be 10′, or 10 nm. The DVL reported that the average errors experienced by three separate observers ranged from 27′ to 49′. Plath did not build a periscopic sextant with an averager before World War II.[48] The periscopic principle was not copied in the English-speaking world until after the war, so the very useful feature for checking compass deviation was not available to them for several years.

If the navigator was not to open a hatch for celestial observations, and in the absence of a periscopic sextant, the aircraft designers had to provide windows or domes for the navigator's use. Aeronautical designers resist protuberances of all kinds. Amelia Earhart and Howard Hughes outfitted their world-flight airplanes lavishly. For celestial navigation, both provided low-distortion glass in various windows. The Hughes airplane also had "a hatch with a demountable windshield," which could be considered a step toward a permanent dome. Even with its emphasis on celestial navigation, Pan American was slow in adding domes to its aircraft. The first two Clippers, the Sikorsky 42 and the Martin 130, provided nothing better than windows and hatches. The third, the Boeing 314, which entered service in 1939, was the first with an astrodome. Curved glass would have introduced unknown refraction for the navigator to contend with; the Boeing solution was therefore a low, teardrop-shaped bubble, made of a number of small, flat panes. It was located on top of the fuselage at the mid-wing position, which was the most stable position in the airplane. The Martin Clipper had included the engineer's station in the same area; Boeing increased the space to provide room for the navigator to use the astrodome.[49] The major producer of flying boats for the United States Navy, Consolidated, went through a similar development. The Archbold Expedition of the American Museum of Natural History, which intended to spend two years in New Guinea, bought a Consolidated PBY-1 and refitted it completely. After flying in it in April 1938, before the airplane left the States, Weems pronounced it "apparently the best-equipped plane in the world today" and commended the "small turret [that could be] opened at the top or opened at the side" for sextant and driftmeter work. Consolidated had just unveiled its newest prototype, the XPBY2Y. The four-engine flying boat had a "retractable navigation turret" in top of the fuselage, at the trailing edge of the wing.[50] Designers of long-range aircraft were finally providing the navigator "a place of his own" for making celestial observations.

After making the observations, the navigator still had to reduce the measured height to a position. The many efforts toward more precise altitude measurement were matched by equal activity directed toward faster reduction. One step favored by all aerial navigators who practiced celestial navigation was the production of an almanac more suited to their needs. The nautical almanacs, published by sev-

The AT-7 navigation trainer in 1942, showing the rotatable celestial turret. *USAF photo.*

eral maritime nations (above, p. 16), gave the locations of the various astronomical bodies used in navigation, but the coordinates were those of astronomers. Further, for the sake of a more compact publication, positions were tabulated only at wide intervals of time, and two systems of time were used, one based on the sun, the other on the stars (see footnote, p. 34 above). The marine navigator could afford the time to interpolate between the tabulated positions and to convert between the different systems of coordinates as well as the different times. Aviators were in more of a hurry.

The *American Nautical Almanac,* lunar supplement, and associated tables, in a canvas bag for convenient cockpit stowage.
Weems & Plath photo.

The first step in the evolution of what would be an air almanac was taken by the U.S. Navy in 1929. Weems, working in the Hydrographic Office, suggested that the moon was not being used by either surface or air navigators as much as the other celestial bodies because, having most erratic motions, it required more interpolations than even the surface navigator was willing to make. Therefore many navigators were passing up the chance for daytime celestial fixes, during those periods when sun and moon were visible simultaneously. To make moon observations more palatable, a supplement to the nautical almanac was published in 1929. For every ten minutes of Greenwich time, the position of the moon was tabulated in declination (equivalent to latitude) and Greenwich hour angle. GHA is similar to longitude, except that it is measured westward from Greenwich from 0 to 360°, while longitude is measured both east and west up to 180°. Tabulating GHA in the almanac eliminated one step in the solution of a moon sight, while the more frequent tabulations simplified what had been an especially difficult interpolation. The moon supplement made "the moon easier to handle than the sun."[51]

Lindbergh's navigation equipment on the 1933 survey flights for Pan American.
Photo courtesy of Capt. Weems.

To extend the same principles to the other celestial bodies, the Naval Observatory and the Hydrographic Office published an experimental air almanac for the year 1933. Weems was again closely involved. The new almanac gave the position of all the navigational bodies in terms of GHA and eliminated sidereal time. Lindbergh used it in his survey flights for Pan American in 1933 and found it especially helpful, but the surface sailors who controlled the Navy were not convinced that the duplication of effort was worth the cost. "Most unfortunately, the Air Almanac was discontinued in 1934, though some of its features in emasculated form [were] included in the 1934 Nautical Almanac." France next took up the project of an air almanac, adding a valuable contribution: all data for one day was printed on the two sides of a single perforated page so that, while the annual publication required 730 pages, the navigator could tear out those pertaining to his flight and leave the rest on the ground. The French continued the use of GHA, tabulating position information for the navigational bodies every 20 minutes of time.[52]

In 1936 Weems, in England in conjunction with the publication of a British edition of his navigation textbook, discussed the need for an air almanac with officials of Hughes & Son, the Admiralty, and the Air Ministry. Both military and civil members of the Air Ministry supported the publication of a British air almanac, and it first appeared in September 1937. Generally it combined the features of the 1933 American almanac with those of the French almanac.[53]

Weems was soon publishing the British almanac under license for sale in the United States, and the Army Air Corps became one of his best customers. The Navy was forced to reconsider the question in 1939.[54] In spite of personality conflicts and worse than usual organizational in-fighting within the bureaucracy, the American air almanac duly appeared.[55] Thurlow, on hearing that the Naval Observatory intended to publish an air almanac, wrote Weems in late May 1940: "Naturally we're happy to have a domestic source (looks like the British won't be putting theirs out in a few weeks; nor needing them either)." Fortunately the British continued to need an air almanac, and the irrationality of duplicate computations on both sides of the Atlantic led after the war to joint publication.[56] Today's *Air Almanac* appears three times yearly, bearing a joint imprint:

Washington:	London:
United States Naval Observatory	Her Majesty's Stationery Office

The almanac provided the location of the celestial body; the sextant measured the height. These items defined the celestial triangle, which still had to be solved. The preferred methods in the early post–World War I years included various marine tables and the Bygrave slide rule (above, p. 111). Two of Britain's most famous aerial navigators, Francis Chichester and Donald C. T. Bennett, relied on the Bygrave rule.[57] But "a large number of tables were published with special reference to the air, and each claiming to be the quickest and most accurate: Ogura, 1920; Newton and Pinto, 1924; Smart and Shearme, 1922; Goodwin, 1926; Weems, 1927; Dreisonstok, Ageton and Gingrich, 1928–31; and Aquino, 1933." The Naval Institute published Weems's tables after he agreed to reimburse them for any loss. He drew heavily on the method of the Japanese astronomer-mathematician Ogura; Ogura in turn had merely applied a formula suggested in 1850 by the first professor of navigation and mathematics at the U.S. Naval Academy. The Dreisonstok and Gingrich tables appeared hard on the heels of Weems's. Weems interpreted the appearance of Hydrographic Office publication number 208 (Dreisonstok) as a backhanded compliment. Gingrich's, published privately, differed primarily in that Weems provided azimuth by a diagram, Gingrich by tabulation. In the midst of this activity, E. B. Collins of the Hydrographic Office began another set of tables but abandoned the effort, observing that "it is difficult at this late date to juggle the astronomical triangle to a new form." Nevertheless, the tables that would be most widely used by American aviators in World War II, H. O. 214, did not start to appear until 1936. Arranged in bands of 10° of latitude per volume, so that a navigator need carry less than the complete set of nine volumes, the first eight volumes, covering 79° South to 79° North, were available by Pearl Harbor.[58]

Generally, any tabular solution of a celestial observation required extracting information from three or four different pages and writing down, adding, or subtracting figures 20 to 25 times, depending on how one counts the "steps" of the solution. Every table had its ardent supporters; it is difficult now not to conclude that the first method a navigator learned tended to be his favorite, and that evaluating one against the other is rather sterile. To the uninitiated, all tables look forbidding. More important, toward the end of a long flight, mistakes are easily made in the simplest computations. On both counts, tabular solutions are undesirable. Again following the direction taken by the balloonists (above, pp. 32–36), a number of nontabular methods of reduction appeared between the wars. They can be classified as graphical and mechanical.[59]

The best-known graphical reduction method to appear between the wars was Weems's star curves. Sets of precomputed curves had been used by balloonists before World War I, but Weems got the idea from Wimperis's *Air Navigation* of 1920 and a National Advisory Committee for Aeronautics report published in 1924.[60] Weems's patent application stressed that, with the curves, less schooling was required of the aerial navigator, fewer items needed to be carried in the aircraft, no assumed position was needed for plotting, and a pilot

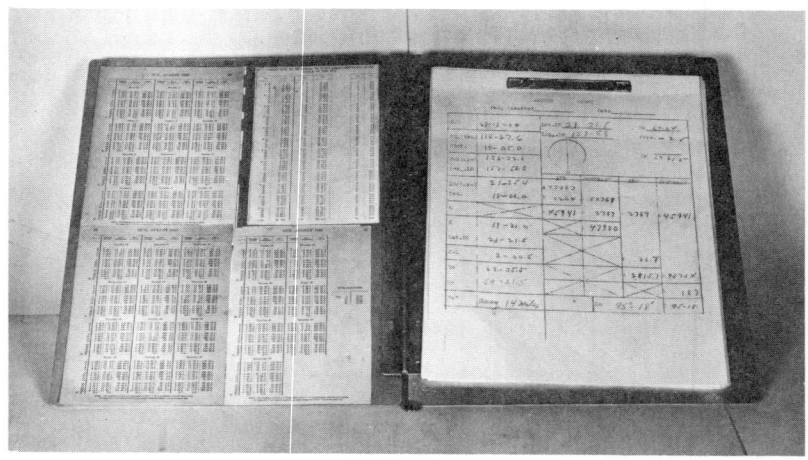

Tables and computations for reducing a celestial sight by the Ageton method.
U.S. Navy photo, courtesy of Capt. Weems.

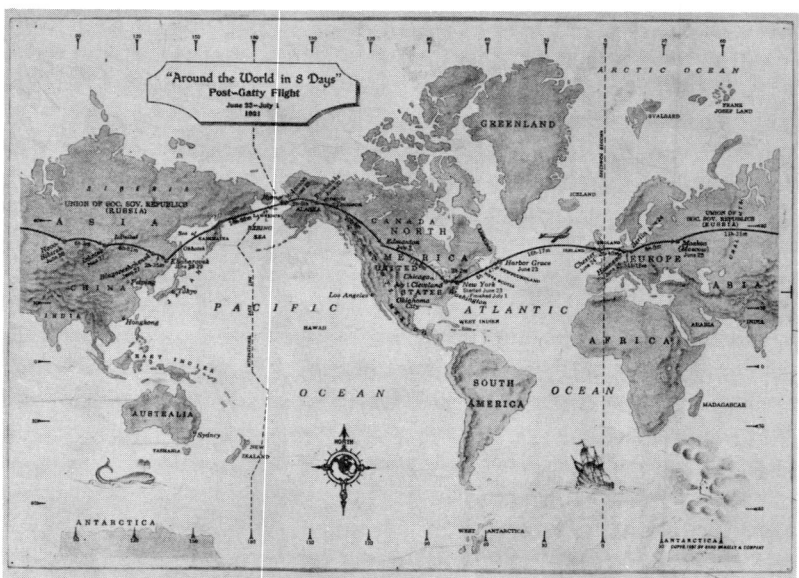

Route of the Post–Gatty world flight, 1931.
Weems & Plath photo.

Navigator's cockpit of the Breguet *Question Mark*, showing pelorus, September 1930. *USAF photo.*

could carry out the whole operation with only part of his attention. Only the last claim could be seriously questioned. While Francis Chichester several times used a hand-held sextant while flying solo, Lindbergh specifically denied being able to do so.[61] Gatty, by then associated with the Weems System of Navigation, used the second edition of the star curves on his world flight with Post and praised them. This testimonial was not the best, not only because of the economic interest, but because after landing in what they thought was Ireland, they found they were at Chester. Gatty's work over Siberia was more impressive. Two major criticisms were leveled at the curves. First, only two stars were provided. Later editions increased this to three, but still clouds over part of the sky might prevent use of the selected stars. Further, when the curves first appeared, navigators were shooting from open cockpits. As their view of the heavens was later more restricted, it was less likely that they could see the particular stars. Second, if celestial navigation had to be used during the day, a different method of reduction was required. A student could learn a single, universal method more easily than two.[62]

A French variation of the same principle, popularized by Bellonte's

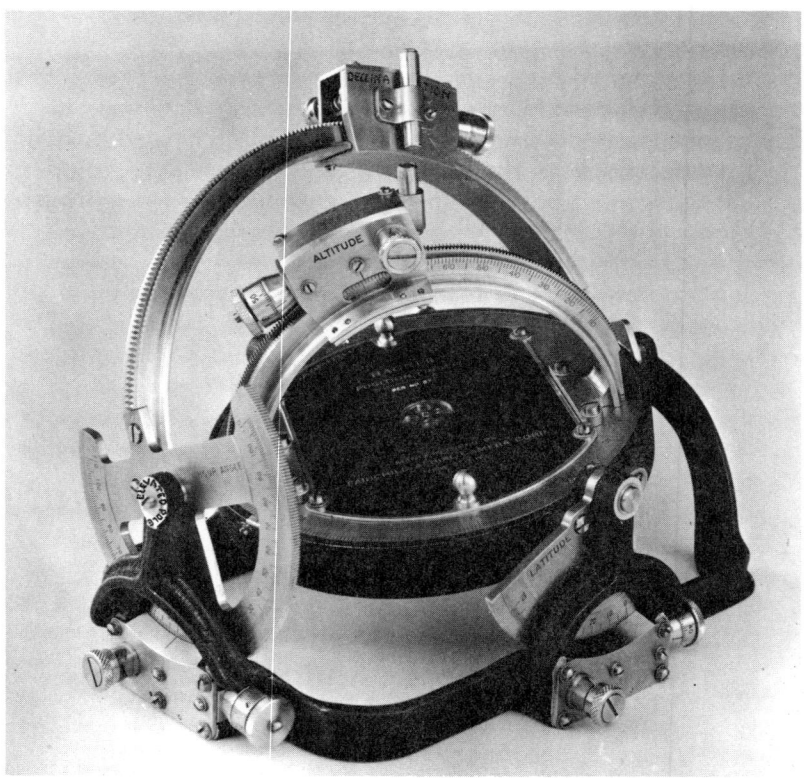

The Hagner position finder, marketed by Fairchild Aerial Camera Corp.
Weems & Plath photo.

successful use of it in 1930, printed the altitude lines on the navigational chart. The chart was long and narrow, not the easiest to handle in an airplane, and only those celestial bodies whose positions could be plotted could be used.[63]

Much more elaborate than sets of curves were a series of mechanical reduction devices. One type replaced the sextant as well as the tables. The bubble level and the sextant optics were attached to a set of three interconnected circles that represented the variables of the celestial triangle. Using the instrument, the navigator solved for two possible positions and connected them on his chart to form the line of position. Capable of high precision and speed, the instrument was bulky, very expensive, and used fine scales that were difficult to read in an airplane.[64] Weems dabbled with the possibilities of an instrument that would track a celestial body continually and give the plane's position on an indicator on the instrument panel, but the attempt was 15 years before its time. The electronic computer would be needed to make such an instrument work. A complicated analog computer for converting height observations to geographical positions was marketed commercially by Fairchild Aerial Camera in 1938. The instrument was again heavy and expensive; worse, its excess of "counters, gear differentials, pin cams and the like" made the instrument almost impossible to keep calibrated.[65] Squadron Leader John F. Griffiths, one of Weems's RAF confidants, wrote after seeing specifications of the Fairchild instrument: "Looks like a wireless set, and costs about as much as a Dusenberg. Someone suggested that they fit a loud speaker to it .... which would tell the pilot which way to go .... A very useful aid to homing after a Guest Night in the mess." Less elaborate mechanical devices appeared in France during the 1930s. One was a development of the Favé and Brill instruments of pre-World War I (above, pp. 34–35); another was similar to the Baker plotting machine used by Alcock and Brown in 1919 (above, p. 109).[66] The only mechanical reduction device to achieve wide acceptance was the astrograph.

In October 1938, when Squadron Leader Griffiths had some leisure after the Munich crisis, he wrote Weems enthusiastically about an idea: the use of a transparent navigation chart over an opaque sheet carrying sets of altitude circles. In other words, he had rediscovered Favé's arrangement of 1906. Weems replied with less enthusiasm: he had considered such a device and abandoned it in favor of the star curves. Griffiths did not drop the idea, however, and ultimately interested his superiors in the instrument, called the astrograph. By that time the design had become more complicated. The curves, on transparent film, were run through a projector above the navigator's work table, so that the lines were projected onto the regular navigation chart. Once aligned properly, the navigator simply traced the proper line of position on the chart and turned off the projector. The difficulty was in the alignment. Because the film and the paper had different coefficients of expansion, a correctly aligned astrograph would be thrown out of adjustment by changing temperatures. Further, only a few stars could be accommodated on the film, so that the same objection leveled at Weems's star curves also applied to the astrograph: the navigator might not be able to see the selected stars that were plotted.[67]

Pan American investigated the various new methods but continued to rely on the basic tables, such as Dreisonstok. Pan American, however, could afford to pick its navigators carefully and train them leisurely. With the expansion of the RAF and the U.S. Army Air Corps before and during the war, training programs were accelerated at the very time that entrance standards were lowered. Griffiths, struggling to get his night bombing squadron combat ready in 1938, complained to Weems that "all these systems were too complicated and lengthy to be taught to the average pilot. . . . whose mathematical education appeared to have ceased when [he] could count up to 10." For such reasons the RAF adopted the star curves and the astrograph, while the U.S. Army Air Corps and Navy both adopted the curves, in spite of the limitations inherent in both devices.[68] Neither nation was satisfied with those expedients, however; in June 1941 Thurlow was busily comparing the latest British air navigation tables with the latest Hydrographic Office publication, recommending a new publication to incorporate the "best features" of each.[69]

The development of the various instruments used for aerial navigation between the World Wars has now been recounted: radio; airspeed meters; compasses from magnetic through directional gyro to magnetic-gyro combinations; driftmeters; pressure and absolute altimeters; sextants.[70] The increasing use of the weather map for planning long-range flights and the various methods of reducing celestial observations have been described. In practice the transocean navigators used all these instruments and techniques.

Dead reckoning continued to be basic. The transoceanic navigator read the compass, altimeter, airspeed meter, and driftmeter periodically, being careful to notice any changes from previous values, and used this data to keep a running account of his craft's position. Where the marine navigator computed the coordinates of a DR position mathematically and then plotted it on the chart (above, pp. 10–11), the aerial navigator had simplified the procedure. He combined the forces affecting the aircraft vectorially, found the course and distance traveled from the previous position, and plotted the DR position on the chart directly, using draftsman's dividers and protractor. The dead reckoning was updated regularly—hourly or less, depending on company policy—by whatever fixing methods were available. The navigator did not use only the most convenient navigational aid; he used all of them. Because any one method might fail, it was prudent to use them all, to have a procedure already established before it became necessary. Because one could not know what was ahead on the route, navigational data from all sources should be gathered and stored in the log for possible later use. Because no method was perfect, the radio bearing should be compared with the celestial line of position, and both with the dead-reckoning data. And the peak of navigational perfection was not the ability to carry on several different navigational techniques simultaneously, but the ability to integrate their results. Sometimes the navigator had to judge between conflicting data, accepting one, rejecting another. Sometimes by combining data from different sources, such as a radio bearing with a sunline, he could plot a definite position, where either alone would have given him only a position line.

The small group of transocean airline navigators was in remarkable agreement. Pan American required celestial observations at least hourly day and night; drift measured with reference to aluminum powder bombs by day and magnesium flares by night; continual watch by its own ground direction-finder stations; supplemented by radio bearings measured with the aircraft equipment. British Airways practice, taught by its own navigation school, was to use

> one method to check the other as a matter of routine, even when one appeared sufficient. . . . radio bearings and astronomical bearings should be taken constantly, whether required at the time or not: if not required at the moment they can always be run up [advanced for time] to make a fix with a later bearing.

A scientist of the French Air Ministry, familiar with Air France practice, said:

> In the present state of navigation every source of information may be inexact at the moment of observation and so air navigation becomes truly an art. The navigator must assess as well as he can the value of a Dead Reckoning, a radio bearing or a star observation.
>
> . . . .
> They interlock and complete each other and give, when they agree, certainty and security.

A German summarized his experience on Lufthansa's Atlantic routes:

> The long-range flier must master all forms of navigation and know the accuracy that can be expected of each. . . . it is a question of the proper coordination of all forms of navigation.[71]

And the efforts succeeded. No transoceanic airliner was lost in the 1930s because of faulty navigation. The many critics of overland airline navigation had nothing but praise for the overwater carriers.

This narrative has now arrived at the second cataclysm of the 20th century. Preparing for it, air forces drew heavily on the experience concentrated in the transoceanic airlines. With the advent of war, many military aviators who had been seconded to the European airlines changed uniforms and cargos. Pan American was soon running a navigation school for the U.S. Army Air Corps, as well as sharing trade secrets on how to squeeze the maximum range from a tank of gasoline. The equipment used by military navigators in the 1930s did not differ appreciably from that used by the airlines. Indeed, it has been impossible to separate military from civil in the development of most of the instruments, as indicated throughout this chapter. But to conclude the story of air navigation before World War II, it is necessary to look more closely at the role of navigation and navigators in military aviation while plans were being drawn for that conflict. As with World War I, the longer-range missions will be of more interest. The equipment and skills that directed a Boeing Clipper into Wake Island could also be used to direct a Boeing B-17 to a designated industrial target.

Or could they?

NAVIGATION OF LONG-RANGE BOMBERS IN THE U.S. ARMY AIR CORPS TO 1941

In peace, armies and navies prepare for war; at least, that is the theory. In preparing themselves for World War II, air forces did not overlook navigation. Naval aviators, especially those operating from aircraft carriers, faced severe navigational problems. Flying from small and mobile bases, they expected to attack small, mobile targets. The targets would be located by individual aircraft flying visual search patterns, after which a strike force would be called in. Because an aircraft's position would be inaccurate after a long period over water, the strike force would home on radio signals from the reconnaissance aircraft.[1] Radio bearings would also be available, if needed, for the return flight to the carrier. Encouraged positively by their naval heritage and negatively by the too obvious penalties of failure, naval aviators paid close attention to navigation.

All air theorists were aware of the possible need to shift their aircraft quickly from one operating area to another. They therefore had to deal with the navigational implications of long flights between fixed bases. Such flights were essentially logistical, however, whether delivering cargo or merely the combat airplane and crew, and the techniques used by navigators on transocean airliners were equally available to military aircrews. This chapter will concentrate on a third branch of combat aviation, one in which the navigational problem, equally crucial, could be solved only partially by the instruments and techniques of the commercial navigator—the long-range bombing mission.

In World War II the Anglo-Saxon nations were the only combatants to commit a major fraction of their national resources to long-range bombardment. The decision to do so was not taken overnight. The United Kingdom had been busily preparing such a force at the end of World War I, and the United States had intended to take part. Between the wars, air theorists in both nations developed the idea of striking deeply into an enemy country, by-passing the tradi-

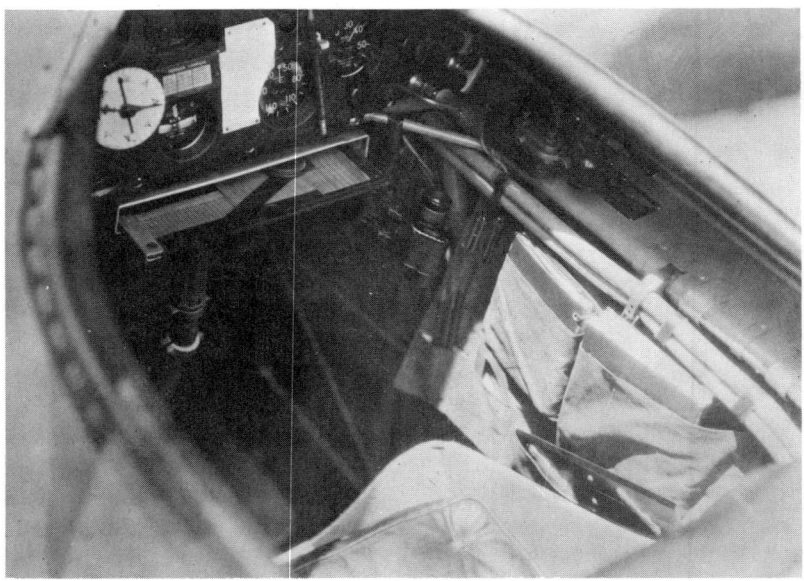

Navigation gear stored in a U.S. Navy O2U, illustrating the cramped conditions long imposed on aerial navigators.
U.S. Navy photo, courtesy of Weems & Plath.

tional military forces to attack the sources of a nation's military power. The idea was elevated into a doctrine. Appropriate aerial vehicles were produced, though never in the quantities deemed necessary by the theorists. Aircrews were trained. A small number of individuals over several years considered the entire problem and experimented with equipment designed to do the job. Yet in the first years of World War II, the long-range forces of Great Britain proved faulty in several respects, one of which was navigation. The United States Army Air Corps was able to use the British experience to a limited extent, but most effort necessarily went to increase the size of the force in the time remaining before Pearl Harbor. In navigation, the Air Corps was essentially no better prepared in 1941 than the RAF had been in 1939. The attention given to navigation by the peacetime RAF, the discovery of the deficiencies in combat, and the steps taken to overcome them have been well described.[2] This account will therefore focus on the Army Air Corps.

To understand the relation of navigation to bombardment at the beginning of World War II, it is necessary to trace briefly the evolution of the doctrine of long-range bombardment as an independent, decisive means of combat. It will then be possible to note the navigational implications of the doctrine as it developed, the extent to which the implications were recognized, and the steps taken to meet the understood needs.

At the end of World War I, German defenses had forced the Allies to rely largely on night missions, although the accuracy of night bombing was dubious (above, pp. 99–100). The two interwar apostles of air power, Giulio Douhet and William Mitchell, gave little attention to navigation. Douhet stressed the freedom of action of the airplane, its ability to fly a straight line in any direction, but said not a word about the solution of the navigational problem implied. In spite of World War I experience, he taught that a bombing force, properly organized, equipped, and indoctrinated, need fear only minimal damage from the defense. He therefore could specify daylight operations, and he asked only that the crews be trained "to drop their bombs as uniformly as possible" over the target, to achieve a geometrical pattern of destruction on the surface of the earth.[3] In his last words on the subject, he said: "An offensive against political, industrial, communication, and other centers does not require great precision in [bomb aiming] in order to achieve terrifying effects." He asked for research to make flying safer; to eliminate from the construction of aircraft those materials that deteriorate in time; to increase speed, range, and carrying capacity; and to decrease fuel consumption.[4] He saw no need for research on navigation instruments.

Mitchell was equally perfunctory toward navigation. While most of his writing was argumentative, directed toward an immediate question, and often dealt with organization, on one occasion he specifically addressed the problem of "how to do it." In about 1922 he wrote "Notes on the Multi-Motored Bombardment Group, Day and Night." It is filled with detailed recommendations about most facets of bombardment operations—including the number of clerks to be assigned to a Service Squadron—but navigation is largely absent. He urged detailed coordination of an attack. "Great care should be taken to arrive at the target at exactly H hour or H hour plus the number of minutes specified in orders." Again:

> There are many methods of approach to the target which should be considered. One is where Bombardment shams an attack on an objective other than that to which it is proceeding, then suddenly turns off to the real point to be attacked.
> Another method of approach would be to proceed toward the target about midway between two military objectives, so that the attack might be launched against either one. . . .
> If there is a large body of water near the objective the

planes will get over this and proceed to a point where they can most advantageously leave it to get at the objective. Where it is possible to go out of sight to land and then come in near the target this is a good way to approach.

Such remarks cry for comment on navigation, but there is none. In spite of the title, there is nothing about night operations. For day flights, Mitchell obviously was assuming navigation by map reading, supplemented by flying precomputed compass headings while over water. He urged research to improve "bomb sights, releases, racks, and related equipment."[5] He, too, saw no need for new navigation instruments.

After Mitchell's court-martial, U.S. Army Air Corps officers abandoned the front page but continued to argue the potential of the air weapon within the Army organization. Because the War Department retained a veto over formal statements of air policy, the best source for following the development of bombardment doctrine is the Air Corps Tactical School. This school, founded at Langley Field in 1920, was moved to Maxwell Field, Montgomery, in 1931, where it continued until 1940. Most key Air Corps officers of World War II attended the school as students, served as instructors, or both. The faculty was closely associated with the Air Corps Board, the body charged with development of doctrine. Remembering the continuing Army Air Corps–Navy struggle and the political impossibility of discussing offensive warfare, the textbooks, lectures, and correspondence of the Tactical School provide a means for following the thoughts of future American air leaders.

In the early post-World War I period, pursuit (the fighter-interceptor) was seen as the most important branch of military aviation, because it was the means to air superiority. In the earliest Tactical School documents, bombardment targets were described imprecisely. Navigation was not seen as especially difficult. "When the distance to the objective warrants it, a navigator for the formation may be announced. He is usually the bomber in the formation leader's plane." Because day operations were expected to suffer prohibitive losses, heavy bombardment would normally operate at night. As early as 1924, however, American bomber crews began to hear a litany that is still sung in some quarters:

> Most important of all in the training of the bombing crews is the inculcation of the will to reach and destroy the objective. Regardless of opposition by the enemy, it must find a way to reach its objective. . . . *This, in fact, is the basic doctrine of bombardment aviation.* The defenses against aviation are numerous; their powers are real. *But no matter how numerous*

or how powerful they may be, they will not prevent bombardment from accomplishing its assigned missions.

It is of course helpful if men engaged in hazardous undertakings can be convinced that they will succeed in their endeavors. More prosaically, the Group Operations Officer would tell the night bomber crew the route to follow "to reach its objective." The bombers proceeded individually, along routes devised to avoid mid-air collisions and timed to bring the bombers over the target in a predetermined sequence. The crews would rely, night or day, on map reading, exactly as in World War I.[6]

The 1931 bombardment course went a little further into navigational requirements. For high altitude, day missions, no problem was anticipated. "The field of view . . . is such that all pilots can be aware constantly of their location. Should the leader be forced down, the mission, so far as navigation is concerned, may be continued without difficulty by the deputy or other pilots." For low-level, night operations, navigation would not be so easy, unless railroads and rivers were conveniently located. Technology promised assistance, however. The radio beacon could be borrowed from the civil air route and set up at the edge of friendly territory, pointing toward the target. The bombers could then fly along the beam, as along an airway. "By employing check points in conjunction with the beacon, very accurate navigation may be insured." One may infer that the authors of the 1931 text were neither knowledgeable nor unduly concerned about navigation, for under the rubric "Navigation Equipment" appeared the following:

> (1) To perform navigation the magnetic and distant reading or earth indicator [inductor] compass will be required for both day and night operations. The visual, or reed type radio beacon indicator and receiving set, may be required. . . .
> (2) Colored navigation lights on the wing tips and empennage . . . .
> (3) Night landing lights installed below the nose . . . .[7]

Pursuit still took first place in Tactical School thinking in 1931, but thereafter the emphasis changed. Instructors started referring to precise targets, deep within the enemy nation, targets chosen to disrupt entire segments of industry. To make such accurate bombing possible, daytime flights were essential. Fortunately for the theorists, the current generation of combat aircraft seemed to allow the bomber to survive in daylight. The B-9 and the B-10 had performances approximately equal to existing fighters; and the B-17, which first flew in July 1935, outperformed all known fighters. The Air Corps also had the necessary instrument to put the bombs on the target: the Norden bombsight. So "the full-blown theory of high-level, daylight, precision bombardment of pinpoint targets" appeared, to remain the Air Corps's primary doctrine through World War II and beyond.[8] The aircrew obviously had to navigate close enough to the target for the bombardier to see it.

To see the aiming point, weather over the target had to be clear. As to weather over the home base and en route to the target, however, the Air Corps had learned to operate under worse conditions than had previously been possible. A Tactical School textbook of 1935 therefore stressed that bomber pilots required extensive training "so planned as to insure excellence in blind flying, avigation,[9] formation flying, and the ability to conduct flights under all types of weather conditions, both during daylight and during darkness." In fact, flying above an overcast would help avoid interception. "Inclement weather, which has heretofore been a menace to aerial operations, especially bombardment operations[,] is rapidly becoming one of its most valuable aids." With the perfection of blind-flying instruments and the autopilot, bombers could soon "proceed to distant objectives by utilizing the cloud formations as security. Thus, one of the great *limitations* of military aviation is rapidly being transformed into one of its greatest *powers*."[10] The text did not expound on the navigational implications of such tactics, but automatic pilots neither navigated nor found targets. Something else was going to be required. In clouds, that could only be radio in 1935. Above clouds, it could only be radio or celestial navigation.

Although daytime operations received the most emphasis, the textbooks continued to deal with night bombardment missions. The 1935 bombardment course stressed the need for accurate navigation at night, to avoid unnecessarily exposing the bomber to antiaircraft fire around targets. That is, the crew had to make an accurate approach to the target so that the bombardier could pick up his aiming point without having to circle the area looking for it. A recent method, full of promise, was the use of parachute flares dropped at regular intervals by observation aircraft. "They constitute definite guides for the navigation of the bombardment airplanes. This method of aiding bombardment to arrive at its objective has been successfully carried out in peace-time maneuvers and exercises." Nothing was said of how the flare-droppers found their way. Contemporary RAF night-bombing theory was exactly parallel; there, too, "how the first aircraft could find the target in order to illuminate it was not explained."[11] When theory gave way to practice and unacceptable numbers of British bombers were shot out of the daytime skies over Europe in

1939 and 1940, the Tactical School increased its emphasis on night operations. A lecture delivered on 18 June 1940 dealt with bombing an air base at night. Flares were very much part of the scenario. A bomber might drop flares to aid subsequent bombers, but this would never suffice for aiming its own bombs—the flare is always behind the airplane that drops it, and the bombardier must see well ahead. Therefore an "illuminating unit" of aircraft had to precede the bomber formation. The instructor also answered the question raised above. On dark nights, "it is quite probable that the first illuminating wave may have to drop flares at wide intervals over long distances, to orient itself into the vicinity of the objective." Little experience was available for dividing the force between bombers and illuminators, or the load between bombs and flares, but "as a general rule, it is advisable to plan to use *lots of flares*." The necessary flare did not then exist, but the Aberdeen Proving Ground was developing a 1,000,000-candlepower flare with a burning time of 3 to 3 1/2 minutes, having a delay fuze that would allow high-altitude release, low-altitude opening and burning. The source of light needed to be as close to the surface as possible.[12] The pendulum seemed to be swinging all the way back to night operations. On the eve of Pearl Harbor, the Chief of the Air Corps wrote for public consumption: "Successful bombing is done largely at night; targets selected for night bombing must be illuminated. . . . Navigation is important; the target must be found before it can be attacked."[13]

The interwar theorists of aerial bombardment recognized some of the implications for navigation and attempted to solve the problems that they recognized. They attempted to foresee the many ramifications of the bombing doctrine, but in the area of navigation, they unfortunately had slight experience on which to base their imaginative constructions. The Army Air Corps officers who were in positions of authority in the 1930s had been trained in World War I or in the 1920s. The navigational experience of American aviators in the war had been minimal. They had drawn on Allied experience in navigation, as in other areas, but the training they believed essential for postwar student aviators was neither lengthy nor rigorous.

The British certainly had a better chance to appreciate the importance of aerial navigation than did the Americans, and the British experience is therefore illuminating. Trenchard was concerned with the problem in 1919; he would not have RAF officers be "chauffeur[s] only. . . . Navigation, meteorology, photography and wireless are primary necessities if the Air Force is to be more than a means of conveyance, and the first two are requisite for safety, even on the chauffeur basis." The RAF rearranged the wartime navigation course, moved it to Calshot, and accepted applications from regular officers beginning in 1920. The "long course," covering the entire field of navigation, was not popular. In mid-1925 the Royal Aeronautical Society was told there was no school of air navigation in England. An RAF pilot in the audience corrected this error and proceeded to make a second. He said there was a Service School at Calshot, but it taught only air pilotage, or map reading. The long course remained on the books, but apparently there were so few students the officer did not know the course existed. In the early 1930s, Donald C. T. Bennett taught the short navigation course at Calshot and took some of the subjects in the long course. In 1934 he passed the First Class Navigator's examination and became the seventh holder of the license.[14] Francis Chichester offered an explanation for what he considered the sorry state of aerial navigation in 1939. The trouble had started in the war, when most pilots flew short-range missions without decent instruments.

> A very fine pilot, who was flying all through the war, told me the other day that he had never used a compass. The best way to get back was to fly in the general direction of home until the pilot recognised some landmark.
> And so a tradition was founded that navigation was a novelty unworthy of a warrior . . . . we [younger pilots] were inclined to imitate their attitude, and anyone who felt that a little navigation might be a bit of help on a long flight crept off to study it in a locked room with the blinds down. This tradition that flying and fighting are the things while navigation is only a fad persists to-day, and has done immense harm to British navigation.[15]

If some allowance is made for Chichester's overstatement, the American experience resembled the British. The Equipment Section of the Air Service's Engineering Division (at McCook Field, Dayton) was responsible for research and development of all aircraft accessories, which included instruments. In 1919 the section tested parachutes, interphone systems, self-sealing gas tanks, and folding funnels and buckets for filling tanks. It was sufficiently interested in the following items to solicit external assistance: air-bag floats, landing skids, gas gauges, engine driven electrical generators, and powered cranking devices. In 1920 a Navigation Branch was organized within the Equipment Section. Thereafter navigation had an institutional home in the command that was charged with research and development, but the priority accorded navigation remained low. Officers assigned to the Engineering Division ordinarily first attended the Air Service Engineering School, also at Dayton. *One week* of the 50-week school was devoted to navigation and meteorology. "All the valuable methods of navigation are taken up in brief summary, and navigation instruments including air sextants are employed."[16] The engineering students were of course pilots and had been exposed to navigation earlier, during pilot training. How extensive had this exposure been?

Once again following the British lead, the U.S. Army at first sent its aviators to a naval school to learn navigation. Later the Army included navigation in its own schools. In 1920 pilot training lasted four months, after which time the fledgling flying officers went to one of the three specialist schools, corresponding to the branches of combat aviation: observation, pursuit, or bombardment. The officers selected for observation first attended the Field Artillery School for three months and then entered the Air Service School of Aerial Observation for three more months. Then, in *one week*, "the classes covered Aerial photography, radio, artillery liaison, aerial navigation, map reading, map making, and visual reconnaissance." Pilot training was concentrated at San Antonio after 1927, and more time was alloted. In the first phase of training, lasting eight months, 24 hours of the 360-hour ground school were devoted to navigation, with 20 additional hours spent on maps. In the advanced phase, during which students specialized in one of the branches of combat aviation, now enlarged to four with the inclusion of attack, none of the ground-school courses included "navigation" in their titles. In 111 hours of advanced flying training, the bombardment student spent 48 hours in day and night navigation.[17] Such figures can be misleading, however, for to this day any flight that departs one base and lands at another is apt to be classified "a navigation training flight," on the grounds that some navigation must be practiced in order to complete the flight. The ground-training curriculum for student pilots was "on the whole remarkably stable" throughout the 1930s. Navigation and maps averaged 37 hours, slightly more than half the time spent on buzzer (code) practice.[18] While it is dangerous to equate classroom or flying hours with the amount of learning absorbed, it is safe to say that Air Corps pilots were only being exposed to navigation. If they were to master the subject, something more was needed.

Quite early—before the spate of transoceanic flights of 1927 aroused interest in navigation, before the civil airways were equipped with any navigation aids more sophisticated than rotating beacons, and before the Air Corps possessed aircraft with performance characteristics significantly superior to those of World War I—a board of officers convened at McCook Field in January 1927 "for the purpose of discussing aids to navigation, particularly instrumental aids." A board of officers, infrequently used today, was a common device before

World War II. It was an ad hoc committee that studied any problem of concern to a commander and made formal recommendations, which were not binding. That a board was convened in this instance shows that someone in command was concerned about the state of navigation. The board, under the presidency of Major Henry H. Arnold, concluded that "considerable of the apparent lack of development of navigating instruments has been due to apathy toward navigation training throughout the Service," and recommended:

> That courses of instruction in the use and operation of the sextant for aerial navigation be inaugurated at the Primary and Advanced Flying School.
>
> That considerable air navigation training be included in the training program of the Tactical Units throughout the Service; such training to be performed over territory unfamiliar to the navigating personnel, utilizing only blank charts with latitude and longitude markings.
>
> . . . .
>
> . . . that the possibilities of the radio beacon interlocking signal visual indicator as an aid to navigation . . . be fully investigated . . . .
>
> That the training schedules . . . for all Tactical Units include in the navigation course at least one lecture by a representative of the Instrument Unit, Materiel Division,—this lecture to be given for the purpose of stimulating interest in navigation and bringing regular personnel in the Service up-to-date in instrument development and to clarify difficult questions arising during the Navigation course . . . .

The next year a navigation school, including some flying training, was organized at Dayton; but after two classes of eight students it was discontinued. The Chief of the Air Corps decided that a separate navigation school was not yet practicable.[19]

All of the Arnold board's recommendations, excepting the first, would be implemented in time, and navigation schools would reappear. Meanwhile, the Tactical School was attempting to make pilots more knowledgeable in navigation.

From at least 1928, the Tactical School included a course in navigation, and the history of that course is strangely representative of the larger story of navigation in the interwar Air Corps. In the first place, its purpose was not to train navigators—there were none such as yet—but to teach pilots some navigation, or, better, to remind them of things they supposedly once had known but had forgotten through lack of use. Second, the course never found an organizational home but was shuttled from one section of the curriculum to another, at times being included with the classroom courses on the theory and application of air power, at other times lumped with "Practical Flying" and administered on the flight line.[20] Finally, the instructors seem to have sensed a special need to justify the course, or navigation itself, to their students.

In 1929 the students were told that navigation was essential for aircraft to reach their objectives in minimum time, for several units to rendezvous, to determine if an objective lay within the radius of action, and to intercept naval craft. In 1934 the instructor, Captain Grandison Gardner, explained the purpose at length during the first class period. Pilots could not fly by landmarks at sea; they would have trouble recognizing landmarks in enemy country. Aircraft had to operate day and night, in all kinds of weather, to the limit of their range, and the ranges were steadily increasing. Military tacticians had to be able to understand the specialists, the nature of technical obstacles and resources. That same year, the flight missions associated with the navigation course were described as intended "to give students a small amount of experience in flying exclusively by dead reckoning," implying that young Air Corps pilots had little or no experience with dead reckoning, which is the basis of all navigation. The students apparently were not convinced, however, and on 20 December 1934 the school addressed a special memorandum to them: "The ability to perform aerial navigation is considered a basic qualification of every airplane pilot and observer and inherent to the qualifications for these ratings. Therefore the Aerial Navigation course at this school should be considered simply—refresher instruction with the opportunity for students to become acquainted with the latest methods and developments and fit them for subsequent aerial missions in the various courses." The students were then asked to complete a questionnaire, "to assist in improving the course."[21]

The responses to the questionnaire were unfavorable, and the next year the instructor, Major Gardner, doubled his previous efforts to justify the course to the students. He pointed out that it was not a navigation school and that they could not expect to become navigators. But all other instruction in the Tactical School took a knowledge of navigation for granted. "We think further that we should say enough about it to convince you, if you are not already convinced, that we can fly when and where we wish if navigation is properly developed." Gardner tied the necessity for navigation to the needs of national defense. He admitted that contemporary defense policy apparently was to rely on the Navy first, and on Army bayonets, artillery, and air power only after the Navy had failed. But:

> If you had some fat hens in a hen house and a fat [man] who loved chicken pie lived near by, you might wish to protect

your chickens from being stolen for chicken pie. Would you provide for keeping this fat [man] out of the hen house or for punching him in the belly with a bayonet after he got in? By all means, try to keep him out as, otherwise, he might have killed some of the chickens before you got there and others might be killed in the fight, and, worst of all, the [man] might win the fight and take all the chickens. So, with our national defense, make provision to keep the enemy out; let the bayonets be the last forlorn hope, opinions of master military minds notwithstanding.

. . . .

. . . . The big need of the Air Force is to prevent an enemy ever landing . . . .

The point I am making is that our wartime mission is over the sea where there are no railroads to follow or over foreign land with which we are unfamiliar and, of which, we may not have the detailed maps we have of our own country.[22]

In 1937 the assistant commandant had to prepare a formal defense of the navigation course. He admitted that it had not necessarily formed "a part of the tactical teaching required in this school," but he believed it had "enhanced the value of both practical and theoretical work in other subjects." Nevertheless, the assistant commandant clearly saw the course as "an expedient due to the lack of basic qualification of airplane pilots sent to this school." "It is possible it may have been responsible for the prevention of injury to personnel and materiel in some cases." If future students were better qualified, the course might be eliminated. Meanwhile, it was reduced from 25 to 19 hours and, in scheduling, "these periods were fitted in where they did not disturb the continuity in other courses."[23] Where Gardner had been trying to arouse interest in the relation of navigation to fundamental air doctrine, the assistant commandant was more concerned with keeping dumb pilots from breaking his airplanes. The navigation course apparently did not arouse much excitement for the remainder of the life of the Tactical School.

Student pilots were taught the rudiments of navigation. Students at the Tactical School got a little more, whether they saw the need for it or not. As the range of the airplanes assigned to the Army Air Corps increased, and as the bombardment doctrine gained ground, there was a growing, general awareness that more was needed. An independent navigation school having been rejected, the alternative was to charge the tactical units with navigation training, which had been one of the Arnold board's recommendations. From 1931, short courses in navigation were taught in various Air Corps units. Generally the courses were one-time offerings, or if repeated at intervals they possessed little institutional permanence. But for about a year, the General Headquarters Air Force bombardment groups at Langley Field and at Rockwell Field, San Diego (later moved to March Field, Riverside), administered navigation courses to which all Air Corps units sent students. The schools were called Advanced Avigation Training Units because "schools" required congressional approval.[24]

The bombardment group courses, which began in the fall of 1933, were by far the most ambitious attempt yet made to teach navigation to Army officers. The courses consisted of 160 hours, half of which were spent in the air, over a period of six to eight weeks. The flying training was primarily over water, and emphasis was divided between dead reckoning and celestial navigation. At both bases, the courses were interrupted for the Army airmail operations in 1934. Then in July 1935 each tactical unit was again charged with teaching navigation to its assigned pilots, and the Langley and March schools no longer received students from other bases.[25] Although the two schools operated only a short time, they are of special interest to this story.

First, the schools were contemporary with the most provocative work in navigation at the Tactical School, described above. It may be assumed that in 1933-35 Air Corps commanders were explicitly concerned with navigation, and the reason is not hard to find. Airplane ranges, especially bomber ranges, were increasing—the B-9 and B-10 were in the active inventory, and at the end of the period the B-17 could be foreseen. Pilots easily equate distance with the need for a navigator on board. Not only were the airplanes flying farther, but the coast defense arguments under which the Air Corps justified its bombers to Congress and the public implied salt water, and a pilot senses the need for navigational assistance very soon after leaving the coast. Having identified an area of weakness in its flight crews, the Air Corps established schools to correct the shortcoming.

The schools were not intended to be permanent, however. Rather, a few individuals were to learn the mysterious skill and then be scattered throughout the Air Corps. Second Lieutenant Curtis E. LeMay attended the first class at Langley (fortuitously: he had been scheduled to attend Communication School; but when the hangar housing that school burned down, his orders were quickly changed). He wrote much later: "I think it was prayed in Washington that we would spread the good news around . . . that little scraps of our advanced training would rub off on the personnel with whom we came in contact." The students were, in fact, expected to establish navigation courses at their home bases.[26] Military officers are often asked to perform difficult feats; but requiring a number of pilots to learn enough

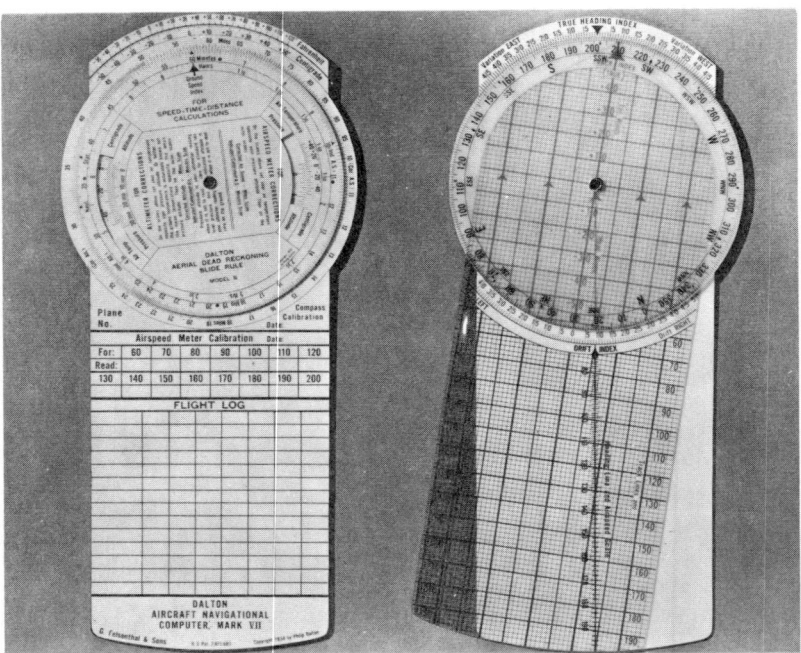

Dalton Mark VII navigation computer, 1934. Students at the Tactical School were not issued such expensive equipment; rather, they "made do" with paper models.
Weems & Plath photo.

about navigation in less than two months to be able to *teach* the subject indicates clearly that the commanders did not appreciate the complexity of the matter.

The two schools had first priority on navigational equipment throughout the Air Corps, but even so, serious equipment shortages developed. Something has been said in the previous chapter of the shortcomings of air sextants, military and civilian, in the 1930s. The two navigation schools found that they could not keep sextants, watches, or bearing compasses in usable condition. During this same period, students at the Tactical School had to use homemade dead-reckoning computers, fashioned out of heavy paper disks held together by metal fasteners. While they were taught to correct indicated airspeed for temperature changes, some of their training aircraft were apparently not equipped with outside air thermometers, because the students were told: "20 degrees C. is comfortable room temperature, and 0 degrees is freezing. The pilot can estimate between these temperatures sufficiently well." Further, instructors at both bombardment-group schools were forcibly reminded of the necessity for painstaking calibration of the various instruments, a task that obviously had too often been overlooked previously.[27]

The need for celestial navigation was also questioned in this period. As with navigation generally, the argument was often stated as a function of range. For short flights, dead reckoning could be surprisingly accurate; but errors in dead reckoning are cumulative. No one expected precise work from celestial observations—as late as 1941 Air Corps combat-crew navigators were required to demonstrate an ability to plot celestial positions that were only within 25 miles of the aircraft's actual position—but a celestial line of position was no more inaccurate late in a mission than just after takeoff. Therefore, if a flight was long enough, at some point celestial navigation became more reliable than dead reckoning. An article in a contemporary French military journal put some numbers into the equation. "It is absurd to speak of celestial navigation for a flight of 150 nm. . . . But if it is a question of trips on the order of 500 nm, it is equally absurd to neglect celestial navigation." Gardner, instructing at the Tactical School, noted the increasing emphasis on flights above the clouds and argued that celestial bodies then might be the only navigational aid available in combat. He urged his students to "recall the old adage about drowning men grasping straws . . . celestial observations may be, at least, a straw in emergencies, and I feel that we ought to count on using it and continue our efforts to improve it."[28]

Finally, experience in teaching the navigation course led instructors (and the wing commander) at March Field to challenge a basic policy of the Air Corps:

> While the navigation methods and techniques practiced in this Unit can be of great assistance to a pilot navigator, it should be thoroughly understood that this course of training is based on the theory that long range missions will be performed by teams consisting of a pilot and another officer whose principal duty is navigation, and that a thorough knowledge of all navigation means available is necessary to be able to select the instruments and methods applicable to a particular situation.[29]

Since World War I, the Air Service and later the Air Corps had been the preserve of the pilot. To keep senior officers who lacked flying experience from transferring into the Air Corps and taking command positions there, the number of nonflying officers was strictly limited, being specified in military appropriations acts; and flying officers were pilots. When airplanes grew to a complexity requiring more than one man, additional pilots were added to the crew to perform the additional duties. The rationale, so far as it was expressed, rested on two assumptions. First, every task on an airplane was related to every other; a bombardier would be a better bombardier if he understood the pilot's problems. Second, in combat and on long missions, it was desirable that one crew member be able to replace another.[30] The Air Corps had the best of company in this practice: the U.S. Navy expected all its line officers to be able to conn the ship. Unexpressed, but likely present, was an awareness of the social cohesion the officer corps gained by maintaining a common, basic distinction: the ability to fly an airplane. Pilots began specializing even before they left flight school, when they were divided between pursuit, attack, observation, and bombardment. It was simply another degree of specialization for the bombardment pilot to concentrate on piloting, bombing, or navigating. Of course, European air forces that operated large airplanes during World War I had been unable to train all flying officers as pilots, and the U.S. Army Air Corps did not expect to be able to do so during wartime. But during peace, the ideal seemed attainable. That possibility rested on another unexpressed assumption: that each specialized skill was sufficiently simple for one man to master several of them in a reasonable length of time; and it was that assumption that the March Field experience called in question.

As long as peace continued, the Air Corps refused to make the change to specialized aircrew members within the officer corps. The compromise that was sought, from about 1935, was the "combat crew." The first commander of the General Headquarters Air Force, General Frank M. Andrews, saw the existing practice of making up crew lists daily, depending on who was available to fly, as extremely inefficient. Military personnel were subject to a host of temporary details, which took them away from their primary assignments, and aviators took their turn along with everyone else. By giving the flying schedule priority over the detail rosters, Andrews was able to insist, successfully, that "only the regularly assigned members of a crew would fly in an airplane[,] that each man would be assigned a particular task, and that all flying in that airplane would be in connection with training of that crew."[31] This step, which did indeed improve flying efficiency, not only complicated the lives of those who drew up schedules; it also gave the flying officers two conflicting ideals to pursue simultaneously.

The flying officer was still required to be a generalist. In 1938, a B-18 airplane commander had to qualify "as a B-18 pilot, as celestial navigator, as expert bomber, and as expert aerial gunner." Not only the aircraft commander was affected:

The requirements of a fully trained combat crew are:
> All officers of the crew to be qualified as responsible pilots or co-pilots, as celestial navigators, as expert bombers, as expert aerial machine gunners, and as aerial camera gun operators.

Bowing to the real world, an escape clause was provided:
> Until such time as a sufficient number of pilots are qualified as navigators, bombers, and gunners, tentative assignments as "acting" airplane commanders may be made without these qualifications.

The enlisted flight engineers and radio operators were exempt; they could honestly specialize.[32] Pilots who flew on the combat crew as pilots had to qualify in the additional positions only once, but the navigators and bombardiers faced a more difficult problem. To be paid, they had to fly as pilots every month. Not surprisingly, some tried to simplify the problem by doing two things at once, for directives to bombing units in the late 1930s were sprinkled with such sentences as: "The flying time listed above will not be done while flying as pilot, but as navigator entirely."[33]

The contradiction was recognized quickly enough. In April 1936, Lieutenant Colonel Tinker, the commander of the 7th Bombardment Group, Hamilton Field, California, wrote to Lieutenant L. S. Kuter, who was teaching at the Tactical School, and criticized the 1935 bombardment course textbook. Kuter thought enough of the letter to have it reproduced for circulation at the school. Tinker admitted the desirability of bomber crew members learning all the jobs on the airplane, but added: "It appears that individual members of the crew must more and more specialize in their particular responsibility." Nonpilot duties had to "be the primary function of one member of the combat crew, and not a secondary one, as indicated" in the text.[34] Ways out of the contradiction were sought in the Office of the Chief of the Air Corps. In December 1937 the Chief of the Plans Section suggested that pilots who had trained in a second skill be excused from maintaining their qualifications as pilots. He did not suggest changing the basic policy of training all flying officers as pilots initially. The next month he went further. Noting that by July the Air Corps would need 240 navigator-bombardiers who did "not necessarily [have to] be rated pilots," he recommended two steps. First, students eliminated from pilot training should be screened, and those who possessed potential as navigators should be trained as such. Second, legislative changes had to be sought, to allow including nonpilots as flying officers. His recommendations were not then implemented. Not until 1941, when the expansion program made it unavoidable, did the Air Corps commission nonpilot flying officers.[35]

Ground school for celestial navigation, Hamilton Field, California, January 1940.
USAF photo.

Between 1935 and 1940 the tactical units—each bombardment group and each reconnaissance squadron—had continued to be charged with training their assigned pilots as navigators, to fill up their own combat crews. Two steps had been taken in 1935 to facilitate this training. First, the General Headquarters Air Force directed each group to provide a navigation staff officer, whose duties were to advise in all matters pertaining to navigation, manage the local navigation school, and supervise the flow of weather data in the command.[36] Second, the 1st Wing's 19th Bombardment Group at March was charged with "the development of the art of Celestial Navigation and its application to long range bombardment." Specifically, the 19th Group was to be concerned with developing a better sextant, adapting the aircraft to the use of the sextant, and applying celestial navigation to operational exercises.[37]

The directive brought a lengthy counterproposal from the Commanding General of the 1st Wing, Brigadier General Arnold. Remembering that March, along with Langley, had just finished running the navigation schools for the entire Air Corps for more than a year, Arnold's reply is revealing. "At the present time so little is known relative to the equipment and material to be used and the method of instruction to be followed that a good portion of a year will be required before a school worthy of the name can be started." The 19th Group had proposed a training program for itself, which Arnold recommended be approved, during which the Group would use its own pilots as students and instructors. "The proper equipment for the training will be ascertained by test, experiment and actual practice. . . . the problems connected with celestial navigation instruction cannot be solved without considerable time being utilized in 'Cut and try' methods . . . ." Citing past supply difficulties, he asked that Materiel be told the importance of the project; and he requested that all attempts to publish a textbook be postponed until after the year had passed. He also asked for new airplanes. They needed to be large enough to accommodate several navigation stations, suitable for overwater flight, and equipped with autopilots, which were essential to celestial observations.[38]

Some part of the 19th Group's proposal was accepted, because Lieutenant Thurlow, who had been the 19th's celestial navigation instructor, wrote Weems that he had been given a year "to carry on experimental and developmental work in celestial." No celestial school for the Air Corps resulted, however, and apparently the project was not continued beyond that time.[39]

Meanwhile, the tactical units started local navigation schools. The training would have been extensive had the instructions of General Headquarters Air Force been carried out, but that was not to be. When there had been only two schools, equipment and qualified instructors had been hard to find. With the number of schools multiplied, the situation could only deteriorate. Further, the requirement to teach navigation tended to get lost among many other requirements that must have seemed at least equally important to the commanders concerned. Lieutenant Curtis LeMay ran a navigation school at Honolulu; he was initially allotted one hour per week. A few years later the 50th Observation Squadron, at Pearl Harbor, reported its officers "in the process of becoming full fledged navigators" at the local school, but the correspondent added wryly: "The practicability of trying to carry on such a school along with squadron duties and squadron training is a point yet to be proven."[40] The Langley school had to ask for an extension of time before starting the first class, to allow "the instructors to reach a state of training somewhat beyond that to which they are expected to bring the students."[41] Some commanders were intent on producing navigators one way or another. In 1941 the 16th Reconnaissance Squadron, Boise, was called to task for declaring pilots competent in dead reckoning and celestial navigation after 30 hours of training. Altogether, navigation training in the tactical units was conducted "in a somewhat haphazard manner."[42]

While the quality of the unit navigation schools was suspect, the quantity of their output was never sufficient. As the 1930s passed, the inventory of bombers gradually increased and, more significantly, a massive expansion became ever more likely. The solution to the navigator problem, offered perennially, was a separate school, preferably located in an area with good flying weather. The reply was invariable: no funds are available or apt to become so.[43] And there the matter rested until the spring of 1940.

By 1940 the long-expected expansion was under way, but with so much to do, again navigation received less than first priority. The Air Corps needed outside help. For years, Pan American Airways had enjoyed a well-earned—and advertised—reputation in aerial navigation. In 1936 an imaginative officer had applied to the Chief of the Air Corps to be detailed for duty with the airline, to absorb some of its experience in long-range flight. The request was disapproved. Another officer was sent on one flight to Manila and back, but nothing else was done. Then, early in 1940, "General Delos Emmons flew to Lisbon in mufti . . . . He became intensely interested in the navigational technique used to guide the clippers and spent most of his first night on the flight deck, absorbing navigation. In the morning, upon sighting the island of Horta, in the Azores, dead ahead, he turned to [the navigator] and expressed his frank amazement." Emmons was convinced that he had found a stopgap solution. The Air Corps conferred with Pan American officials in April and June, negotiated a contract in July, and the first navigation school to train aviation cadets began on 10 August 1940, at Pan American's Coral Gables base, Florida.[44]

Before any enthusiast for "business know-how" waxes lyrical for Pan American's fast action after years of military-bureaucratic bungling, it must be said that the school did not work out very well. It could scarcely have happened otherwise. Little preparation time was available. The Air Corps officer to command the military side of the operation, Captain Norris B. Harbold, received his formal instructions on 5 August, only five days before the school began. The University of Miami provided logistic support, but the president of the university "had a strong feeling of honor, and stated that 'no contract was necessary between him and his government, and if government [army] representatives insisted on a contract they could quarter their cadets elsewhere.' This situation continued to be a source of trouble." For its part, Pan American's training aircraft were slow amphibians, long retired from passenger service. There were so few aircraft that, to get the specified 50 hours of flight training in the 12 weeks (later, 15 weeks) allotted the course, too many cadets were crowded onto each flight. Further, the students were found deficient in mathematics, and there was certainly no time for remedial training. Nevertheless, at the appointed time the first students were graduated, but the Air Corps did not know exactly what to do with them. The cadets had been told they would be commissioned, but the requisite enabling legislation and implementing regulations were not ready until the following year.[45]

These first navigators, in the sense of being *only* navigators, met an odd reception when they reached the tactical units. Some pilots "issued orders that they were going to guide the aircraft to its destination, and would not in any way follow the navigator's course recommendations." Some of the graduates wrote home that "no equipment was available in the field. Aerial navigators were spending their time sweeping barracks. The pilots were reluctant to accept this new airman. They'd never heard of a navigator."[46] Considering the deficiencies of the Pan American school, there was probably basis for skepticism. In any event, many pilots would have resented the new men who were doubly outsiders: they were not pilots and they had not been trained in a military environment, yet they aspired to the privileges of commissioned *and* flying officers.

The Pan American school was a stopgap, better than nothing while the Air Corps set up regular navigation schools. By November 1940 a small navigation school was under way at Barksdale Field, Shreveport. Those were hectic times. A school would be opened, only to find that increasing pilot training activities there interfered with the navigation school, which had to move to a new base. A staff would be put together just in time to have to split it two or three ways, to provide cadres to launch new schools. Many a graduate found he was immediately assigned to teach his successors. But order gradually emerged from chaos. A curriculum was hammered out and gradually lengthened, from 430 hours in 10 weeks to 714 hours in 18 weeks. The individual schools were at first free to do largely as they pleased; later, as complaints came in from the using agencies, central control tightened and the various schools were standardized. An airplane suitable for navigation training was found. The AT-7, with a rotatable celestial dome, accommodated three students plus an instructor. By Pearl Harbor, three Air Corps schools (at Kelly Field, San Antonio; Mather Field, Sacramento; and Turner Field, Albany, Georgia) were producing navigators (and the Pan American school was being filled by RAF cadets). Together, they had graduated more than 500 navigators for the U.S. Army Air Corps.[47]

While the navigation schools were getting started, the Air Corps discovered that the qualifications needed for navigator cadets differed

The "Navitrainer" at Barksdale Field in February 1941. The student has a driftmeter, air thermometer, airspeed meter, altimeter, clock, and radio compass indicator.
USAF photo.

Multi-purpose navigation trainer built by Link Aviation during World War II.
Link Aviation photo, courtesy of Weems & Plath.

NAVIGATION OF LONG-RANGE BOMBERS

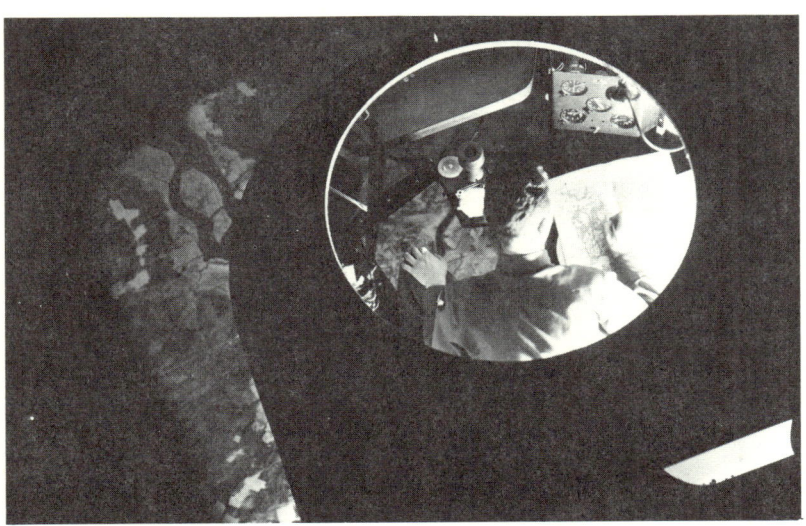

A cadet working a map-reading problem in the Link navigation trainer at Hondo Army Air Field, Texas, in 1944. The map is a sectional of Norway.
USAF photo.

A cadet shooting a star in the Link navigation trainer at Hondo Army Air Field in 1944.
USAF photo.

Students boarding AT-7 at Kelly Field, 1942.
USAF photo.

Interior of the AT-7.
USAF photo.

from those that had evolved over the years for pilot cadets. Flying an airplane is essentially an athletic skill, demanding acute reflexes, sight, hearing, and coordination. It requires greater intelligence as the aircraft systems and the mission become more complex, and as contact or visual flight gives way to instrument flight. No matter how demanding the intellectual requirements become, the pilot whose physical condition deteriorates is dangerous. By comparison, the navigator's work is more contemplative, with less need for instantaneous response. Given his working conditions, which never approach those of the professor's study, an ability to isolate himself from much of his environment is a distinct asset. These differences were anticipated to a large degree before the first student reported to Coral Gables in August 1940.

In July Arnold requested, and received, authority to lower the physical standards and increase the educational requirements for navigator trainees. As noted above, however, Pan American's first students were deficient in mathematics. Through mid-1941, most of the student navigators were cadets who had been eliminated from pilot training and they continued to be, as a group, deficient in mathematics. The impossibility of interpreting transcripts from different colleges led the Air Corps to devise a battery of aptitude tests that have been used, in one form or another, ever since. Throughout the war higher scores were required for navigator cadets than for pilots, although the aptitude scores required of all cadets were generally raised as the war went on, as the following table shows.[48]

Date	Navigator	Bombardier	Pilot
July 1942	5	1	1
Dec. 1942	5	3	3
July 1943	6	6	3
Aug. 1943	6	6	4
Nov. 1943	7	5	5
Feb. 1944	6	5	5

Thus, by the time the United States entered World War II, the Air Corps (renamed Army Air Forces in 1941) had rationalized the duties of the bomber crew and was producing specialist navigators. Yet the AAF was to be plagued by navigational errors during the war. To large extent this is explained by the obvious fact that the earliest navigators were not trained very well and their average experience, on entering combat, was meager. Of more importance, however, is the kind of navigation practiced by the Air Corps before the war.

Perhaps the most spectacular flights made by the U.S. Army Air Corps in the late 1930s were the bomber flights to Latin America. In February 1937 a formation of nine B-10Bs flew from Langley to the

Canal Zone, refueling at Miami. Only the three flight commanders' airplanes had navigators on board; they navigated by sunlines and driftmeter readings, as cloud conditions above and below their flight level permitted. One year later the inauguration of the Argentine president provided an excuse to demonstrate the long-range capability of the B-17. Six of the airplanes—half of the B-17s then possessed by operational units—flew from Langley Field to Buenos Aires, refueling at Miami and Lima. Each airplane carried a navigator and, to achieve maximum training, the airplanes flew individually, joining up near each destination to make a more impressive appearance. One B-17 had been equipped with the new gyro-stabilized driftmeter. The following year B-17s visited Bogotá. In 1939 they went to Rio and a B-15 carried a load of emergency supplies to Chile, following an earthquake there. Propaganda value aside, the flights offered valuable training. The crews had never flown such distances over unfamiliar territory. They had to work with sketchy weather information and rudimentary maps. They discovered that hypoxia was not confined to extreme altitudes; after several hours at 12,000 feet, navigators made ridiculous errors unless they used oxygen.[49]

The bomber crews overcame these obstacles and made good their intended itineraries without so much as scratching a single airplane. Weems talked to several of the officers at Langley after the Buenos Aires flight and wrote to one of his RAF correspondents: ". . . they certainly are a cocky lot. From top to bottom, they have every assurance that they can take off in the so-called 'Flying Fortresses' and go anywhere within the three thousand mile range of the planes with absolutely no hesitancy about the navigation."[50] The crews could well be proud of their accomplishments, but they had really done nothing more than Pan American was doing regularly. The flights had been logistic operations, moving airplanes and crews over long distances, navigating by sextant and driftmeter, with occasional radio-compass bearings. They had not been bombing training missions. At the end of a long navigation leg, when a bombing mission would have been concerned with locating an initial point and starting a bomb run into the target, the aircraft had circled over a rendezvous point, formed up, and then entertained the crowd. The South American flights were excellent training as far as they went. Translated to the milieu of World War II, the flights demonstrated an ability to ferry bombers from the United States to the United Kingdom. The crews had not shown that they could fly from England and attack targets in Germany.

A second category of flights by Air Corps bombers in the late 1930s, which received only slightly less publicity, was more realistic

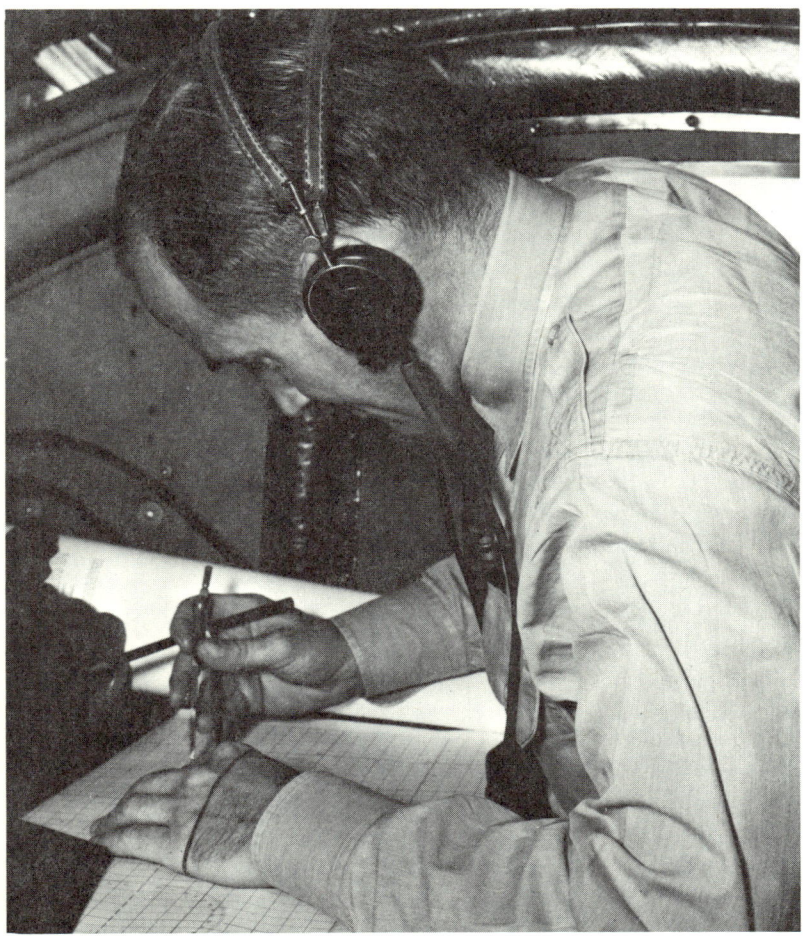

Navigator at his station in a B-17 on the Rio de Janeiro flight, November 1939.
USAF photo.

training for combat. In 1937 B-17s participated in a joint Army-Navy air exercise off the West Coast. On the second attempt the bombers found the *Utah* and hit the buttoned-up battleship with practice bombs filled with water. The next spring, in conjunction with the 1938 war games, the B-17's range was again brought to public attention at a time when, not coincidentally, a naval appropriations bill was moving through Congress. With the cooperation of the steamship company, three B-17s with newspaper and radio reporters on board successfully intercepted the liner *Rex* over 600 miles east of New York City. The event was reported live, on radio. Bombers based in Hawaii used the Army transport *Republic* for interception practice with some regularity.[51] Such exercises were of military value, although they were more appropriate for light bombardment. B-17s were rarely used for such tactics in World War II. They would not seek targets in the Ruhr while flying line abreast, spaced to the limit of visibility, nor would they often need to fly expanding search patterns seeking a mobile target.

The bomber crews flew a third kind of mission directly related to their wartime mission. Many hours were logged dropping practice bombs on bombing ranges. The two busiest ranges were on the Virginia beach and in the Mohave Desert. The target was a pattern on the sand, sometimes the outline of a battleship. The aircraft flew back and forth over the target, ordinarily dropping one bomb on each pass. As the bombardier evaluated each shot and corrected for his errors, subsequent bombs tended to hit closer to the target. Only seldom did a training flight simulate a complete mission. Major General Oscar Westover described one such in January 1937, in praising the Air Corps's all-weather capability.

> A short time ago a crew of the 19th Bombardment Group from Rockwell Field took off from that place with a 300-ft. ceiling, climbed up through the fog, reached an altitude of about 1,200 ft. in the clear, flew above the clouds to the gunnery range at Muroc, about 175 miles away. They came into good weather over the desert, and dropped their bombs on the targets. They then returned to San Diego by using the radio beam, dropped down through the clouds and made a successful landing at their home station. In the future such operations will be the rule rather than the exception. . . . In other words, extended flights will be the rule . . . .

The Air Corps had partially overcome the weather in two phases of the bombing mission—takeoff and recovery. The navigator on the flight described above presumably directed the aircraft's progress to and from the range, but he was flying a well-known route with a weather forecast based on data measured near the target; there were radio stations available if needed; the skies over the target were clear; and the target bore no resemblance to an industrial city. The activity over the range bore no similarity whatsoever to combat conditions, yet bombing scores were compiled from such procedures. When those scores were later used to estimate the probability that a bomber could destroy an industrial target of similar dimensions, a great deal of imagination was needed to adjust for the different conditions.[52]

On at least one occasion, the Air Corps grappled directly with the implications of bad weather over the target. A navigator of the 19th Bombardment Group took a precise fix 50 miles from the bombing range and then navigated by dead reckoning to the target. "Bombsight showed that bomb would have hit about 3/4 of a mile from the point selected, and for a very large target it is seen that this method is feasible. However . . . a very large amount of instrument refinement is required to reduce the error in navigation to a 1/4 of a mile or less." The same unit also experimented briefly with night bombing: ". . . in May, 1938, one night tracking mission was made with a 'blacked-out' objective. No difficulty was encountered in placing the sight on the target, as the blacked-out area could be distinctly picked out from the surrounding roads and miscellaneous lighted objects." A much more realistic night bombing exercise the same year produced altogether different results. During the Air Corps maneuvers of 1938, the town of Farmingdale, Long Island, which was the home of both the Grumman and Seversky aircraft factories, volunteered to participate. The residents blacked out the area for the night of the practice bombing attack.

> "We couldn't see anything," one pilot said later. "It might just as well have been a wheat field at midnight in Texas. We thought that when the residents of Farmingdale turned out the lights we would be able to find the city, as we expected that the other nearby towns would be well lighted. But . . . everyone for miles around had his lights off."
> . . . the pilots of the bombers dropped flares. These turned out to be a help to the defending Pursuit units . . . . No bombs were dropped, of course, the pilots agreeing that had the aircraft factories been an actual target, it would have been difficult to know when to do their bombing.[53]

This may have been the only time before the war when bomber crews were able to practice against a blacked-out industrial region. In 1938 the Air Corps probably interpreted the results as further proof of the necessity for daylight bombing. Whether anyone remembered the Farmingdale experiment three years later, when the British experience was causing the Air Corps to have second thoughts temporarily and General Arnold was telling the public, "successful bombing is done largely at night," the documents do not say.

The peacetime training of Air Corps bomber crews only occasionally approached the conditions of war. For a number of obvious reasons, combat conditions can never be reproduced for the sake of training. Military commanders must approach realism as best they can and then extrapolate from what they can demonstrate in peacetime to what may happen in war. In the case of navigation, that extrapolation was not well made. An RAF officer, one of the few expert navigators with which the British entered the war, wrote in January 1940: "A lot of things which work out quite well in peace time do not turn out quite so well on active service conditions. There is no doubt whatever in my own mind that war and peace require two entirely different methods of getting about the country."[54] When their time came, Air Corps navigators would agree. They entered the war with driftmeter, radio compass, and sextant; before four years had passed, they would be manipulating electronic gadgets that revolutionized aerial navigation. The new equipment was developed during the war because it was essential, if the bombers were to find their targets. The prewar methods of navigation failed. Why was this failure not foreseen and corrected? Or, why was the tempo of change in navigation during the 1930s too slow to provide for wartime needs?

A partial answer is found in the pattern by which air navigation had developed from its beginning. Almost all the instruments and techniques used in the air were borrowed from the sea. Map reading was equivalent to coastal pilotage. Dead reckoning and celestial navigation were taken from marine practice, with such changes as were necessary to simplify and speed up the operations. The magnetic compass and the sextant had long naval histories; the instruments had to be adapted to the high accelerations and instability of the aerial environment. The airspeed meter used the principle of the naval pressure log. Among the early aerial instruments, only altimeters and driftmeters were wholly new. The gyro compass and directional radio were used on ships before the instruments could be made small and light enough for use in the air. The mariner had marine charts; the argument was used to justify special charts for airmen. Buoys and lighthouses, maintained at government expense, were the forerunners of the airways beacons and radio ranges, also maintained at government expense. International cooperation in the regulation of maritime navigation was the model for international cooperation in aerial navigation. The pattern is overwhelming. Unfortunately for longrange bombardment, navies had not devoted much attention to seeing through clouds or at night. Nevertheless, the pattern held. The United States Naval Research Laboratory pioneered research in radar years before the Army took it up seriously, and the first sets, installed on capital ships in the late 1930s, were too bulky for installation on aircraft. The Navy also pioneered loran, another electronic aid to navigation much used by aircraft before the war was over.[55]

Part of the answer is found in the commercial potential of navigation instruments. A number of manufacturers competed actively in the instruments that had immediate commercial application. The airspeed meter, pressure altimeter, and radio compass offered lucrative markets, and competition improved these instruments. Once the airlines became interested in an absolute altimeter, a variety of models appeared. The air sextant, however, found few American manufacturers until the late 1930s, when military demand reinforced the thin market offered by Pan American Airways. No commercial company would commit the large investment that was ultimately required to develop the electronic devices, because no ready market appeared. A commentator at a meeting of American aeronautical engineers in 1929 called specific attention to the necessity for governmental support, if certain kinds of aeronautical equipment were to be developed: "From a military standpoint more attention should be paid to so-called absolute instruments, based on either magnetic or gyroscopic principles. Military aircraft obviously cannot depend for successful operation or navigation upon an easily confused and vulnerable ground system of beacons, light or radio signals, etc." Aerial navigators contributed a few radical ideas for new instruments. An inertial navigation instrument was patented in Great Britain in 1914, but the inventor built it of weights and springs, and the errors were excessive. Inertial navigation had to await developments in electronics; a practical instrument appeared only in the 1950s. Both Gatty and Byrd predicted something akin to inertial, but they had no recipes for building the instruments.[56] Another American navigator voiced his discontent with available instruments in 1936: "We must invent an instrument along quite different lines of a bubble sextant, along a line so original that no one has as yet even thought of it."[57] He was right, of course; but he did not contribute much to the solution of the problem.

A few saw the need for a significant advance in aerial navigation before the war. The usual sources of advance—the navy and industry —did not in this case provide a solution that military aviation could adapt or purchase. And the Air Corps made no effective move to fill the need. Those who commanded the Air Corps on the eve of World War II had received slight training in navigation. LeMay was a member of the first navigation class at Langley, but his generation did not come into command positions until during the war. Lacking an intimate acquaintance with navigation, the commanders of the 1930s not only failed to pursue the radical suggestions; they only slowly allowed changes that were related directly to existing instruments. The continued pleas for giving the navigator better working conditions in military aircraft long went unheeded. The aerial navigators had trouble getting a separate air almanac. The astrodome was similarly delayed. An RAF officer wrote in 1938: "The great trouble has been to obtain the active interest of senior officers in any matter connected with navigation."[58]

A military commander is not expected to be an expert in every skill needed by his command; the staff, properly constituted, makes up for this lack. While bombardment groups possessed navigation staff officers after 1935 and the Materiel Division had a Navigation Branch after 1920, the individuals assigned to these positions seem to have been fully occupied with improving existing instruments and training navigators to use them. Yet the scientific knowledge on which the extraordinary weapons development of World War II was based had been accumulated before the war. What was lacking, at least until 1940, was organization, a means to put those with a military need into communication with those who had an applicable technique.[59]

Lacking personal awareness of the intricacies of navigation, lacking an adequate organization that would put scientists into contact with military problems, the air commanders gave navigation a low priority. According to a 1937 planning document, of the six items to receive priority for research in fiscal year 1939, number six was "Improvement of air navigation equipment." This was not altogether unjustified. The first ingredients for a bombardment campaign were airframes, engines, and pilots. Without these things, there was nothing; but because similar attention was not paid to "complementary facilities," as General Arnold later called them, "America faced World War II relatively ignorant of basic meteorological phenomena and lacking any well-developed knowledge of the air navigational techniques, devices, and facilities which would be essential to sustained, large-scale operation." The responsible commanders had faced official opposition for much of the prewar period and were forced to camouflage their plans. How much time and effort, of the finite quantity available, they wasted on subterfuges can scarcely be guessed, but certainly such time would have been better spent grappling with the real problems. Nevertheless, this small group of men produced the vehicles and the drivers, no small achievement. What they did not do was to foresee the many ramifications of a long-range bombing campaign and provide for all its many aspects, including navigation.[60]

Prediction is a hazardous undertaking, for all that it is essential and commonplace. And even accurate prediction is not enough. The big problems of the future are usually foreseen by many, long before the problems arrive. What is usually lacking is "the proper formulation and enhancement of [the] natural and simple forecast, and an insistence compelling action."[61] The air commanders knew that a

B-17 crew could find a precise target, given clear weather and daylight operations. They believed, rightly, that a bombardment campaign of sufficient intensity could produce significant military results. They also knew what flying weather over central Europe is like. The World War I experience was unambiguous. Commercial aviation had compiled additional meteorological data between the wars. The commanders did not see all of those facts and assumptions together, and no one pointed with "an insistence compelling action" to the central contradiction: that there would not be enough clear days to execute the massive campaign, that weather forecasts would be fallible, and that bombing through clouds and industrial haze would be necessary.

So the Air Corps built a weapon but not a weapon system. The latter term is of course anachronistic applied to the 1930s. Its popularity today indicates a wider awareness of the need for careful, inclusive prediction. One may hope that contemporary predictions—military, technological, political—will prove to be more accurate. If so, civilization may survive, and historians will continue to use their not quite 20-20 hindsight in writing about it.

APPENDIXES

Appendix A

GLOSSARY OF ABBREVIATIONS AND TECHNICAL TERMS

AFArc. U.S. Air Force Archives, Maxwell Air Force Base, Montgomery, Alabama 36112.

airspeed. The speed of an aircraft relative to the air (not the ground). The uncorrected value read from the indicator is *indicated airspeed*. Correction for a number of errors gives *true airspeed*.

almanac. Nautical and air almanacs, more properly ephemerides, give the location of the navigational bodies at frequent intervals of time, and they also include various tables needed in celestial navigation.

altimeter. An instrument for measuring the altitude of an aircraft. The *pressure altimeter* measures the difference between the outside atmospheric pressure and whatever pressure level is set on the indicator. The *absolute altimeter* measures the height of the aircraft above the terrain.

aneroid. An elastic container from which the air has been evacuated, used in the pressure altimeter.

aperiodic compass. A magnetic compass so designed that, after a turn, the needle does not swing beyond the new heading of the aircraft.

assumed position. In celestial navigation, a position near the actual position of the aircraft, for which the height of the body is computed, and from which the intercept (q.v.) is plotted. Depending on the method of reduction being used, the choice of the assumed position may simplify the computation. For example, using a full degree of latitude may eliminate an interpolation.

astrodome. A transparent bubble mounted in the top of the aircraft fuselage, for taking celestial observations.

azimuth. Direction, angle, or bearing from the observer to a specified object.

bearing. The direction of one point from another. The directional reference may be stated. *True bearing* is the angle from true north to the specified object, as viewed by the observer. *Indicated bearing* is the horizontal angle from the nose of the aircraft, measured clockwise.

bearing compass. A compass equipped with sighting vanes, so that the bearing to an object may be determined visually.

compass rose. A circle graduated clockwise from 0 to 360°.

compensation. Any method used to remove deviation (q.v.) from a magnetic compass.

Appendix A

contact flight. Flight in which the aircraft attitude is controlled by visual reference to the earth; opposed to instrument flight.

coriolis error. The error in an airborne celestial observation resulting from the deflection of the liquid in the bubble chamber of the sextant, which in turn is caused by the aircraft's moving over a spinning sphere.

course. The direction of intended flight over the earth; the line connecting departure with destination on the chart. *True course* is measured from true north.

dead reckoning. Determining position by applying speed and direction to a previous position.

dead-reckoning computer. A hand-held analog computer used to solve the wind triangle (q.v.). A circular slide rule and specialized scales are usually included.

declination. The angular distance of a celestial body north or south of the celestial equator, measured from 0 to 90°.

deviation. Error in a magnetic compass caused by magnetism within the aircraft.

DF. Direction finding, q.v.

dip. The angle between the plane of the earth's magnetic field and the horizon is called *magnetic dip*. In celestial navigation, dip is the error caused, when using the natural horizon, by the observer's eye being above sea level.

direction finding. The use of a directional radio antenna to determine the bearing of a transmitter.

directional gyro. A gyro with horizontal spin axis. It holds a constant direction and may be used as a steering reference.

DR. Dead reckoning, q.v.

drift. The lateral effect of the wind on an aircraft, usually expressed in degrees right or left of the heading.

driftmeter. An instrument for measuring drift by visual reference to the apparent motion of the earth. It can be used to determine groundspeed by timing the transit of an object between two etched lines in the optics.

ETA. Estimated or predicted time of arrival.

fix. The geographic position of an aircraft for a specified time, computed by navigational procedures. Thus map-reading fix, celestial fix, radio fix. A DR position is not a fix.

GHA. Greenwich hour angle, the angle measured westward from the prime meridian to a celestial body, from 0 to 360°.

gyro, gyroscope. A spinning mass (a toy top, a wheel, or other form) whose axis tends to remain pointing in a constant direction with reference to space (not with reference to any point on the earth).

gyro compass. A combination magnetic-gyroscopic compass. The gyro provides stability during periods of erratic magnetic signals, as during turns.

Appendix A
205

heading. The direction of the nose, or longitudinal axis, of an aircraft measured clockwise through 360° from true north is *true heading;* from magnetic north, *magnetic heading.*

inductor compass. A remote compass (q.v.) in which a weak electrical generator uses the lines of force of the earth's magnetic field to generate directional signals; popular in the 1920s.

inertial navigation. A self-contained electronic navigation instrument. The output of accelerometers, integrated for time, provides first speed, then distance, which, applied in the direction sensed by gyros, moves counters that show the current position continually.

initial point. In bombing, a landmark (visual or radar) used as a reference point for the start of a run to a target.

intercept. In celestial navigation, the difference between the computed height of a star (based on an assumed position) and the observed height.

knots. Nautical miles per hour.

line of position (LOP). The locus of all possible positions of an observer at a given time. A visual bearing is a straight LOP. A celestial LOP is a circle. In loran, the LOP is a hyperbola.

log. An instrument to measure a ship's speed through the water. The *chip log* was a sea anchor; the amount of line it pulled overboard in a fixed time was a measure of speed. The *patent log* rotated on the end of a flexible shaft; the number of turns per minute could be converted to speed. In the *pressure log,* the height of a column of water indicated speed. By extension, the book in which speed and other data were recorded came to be called the log.

lubber (or lubber's) line. A reference mark on a compass representing the nose, or longitudinal axis, of the aircraft, and against which heading is read.

marker beacon. A low-power directional radio, beamed vertically, used for precise positioning of an aircraft along an airway or the approach to an airport.

most probable position (MPP). The position that the navigator selects as most likely to be correct when an element of uncertainty exists in the navigational data.

NACA. National Advisory Committee for Aeronautics.

night effect. An error in directional radio caused by waves reflected from the ionosphere arriving at the receiver out of phase with the waves that have traveled the direct route.

nm. Nautical mile, 6,080 feet. It is the preferred distance unit in navigation because it equals 1/60 of a degree of latitude, hence it is easily determined on any chart.

pilotage. Navigating by reference to landmarks.

pitot tube. An open tube pointed into the airstream, used in measuring airspeed.

Appendix A

- radio beam. A directional concentration of radio energy. Popularly, the equisignal zone created by the merger of two signals and used to mark an airway.
- radio compass. The Telefunken Company's directional transmitter, dating from pre-World War I, was called a radio compass. From the 1930s, radio compass has meant an airborne directional receiver with an azimuth dial for indicating, more or less automatically, the direction to a radio transmitter.
- radio-range station. A directional transmitter on the ground used to define the airways by generating equisignal zones, or radio beams.
- R.A.E. Royal Aircraft Establishment, Farnborough. Formerly Royal Aircraft Factory, the name was changed to preclude confusion of the abbreviation with that of the Royal Air Force.
- rate. The average daily error of a timepiece.
- remote compass. A compass in which the magnetic element is placed in the wing tip or tail, to be as far from magnetic disturbances as possible.
- running fix. A position computed by two or more bearings taken on a single landmark and adjusted to a common time.
- sextant. An instrument for measuring angles. The marine sextant measures the angle between the natural horizon and a celestial body. The air sextant incorporates an artificial horizon, usually an air bubble. Similar to an octant.
- sidereal time. Time computed by reference to the fixed stars. Ordinary time is computed from the sun.
- Sumner line. A line of position (q.v.) found from observing the height of a celestial body.
- sunline. A celestial line of position based on observations of the height of the sun.
- track. The actual path of the aircraft across the ground. The intended path is *course*. The RAF and USN do not observe the differentiation between these terms.
- variation. The angle measured from true north either east or west to magnetic north. A compass error caused by the magnetic pole being displaced from the true pole.
- wind triangle. The vector triangle composed of (1) heading and airspeed, (2) wind direction and velocity, and (3) track and groundspeed.
- WML. Weems Memorial Library, Division of Naval History, Smithsonian Institution, Washington, D.C. 20560.
- WRes. Residence of Captain Philip Van Horn Weems (USN, Retired), Randall House, Annapolis, Maryland 21401.

Appendix B

INDIVIDUAL FLIGHTS REFERRED TO IN CHAPTER 6

Legend: P, pilot; N, navigator; R, radio

Takeoff Date	Crew
17 Mar. 1924	Frederick L. Martin + 7
12 Oct. 1924	Hugo Eckener + 32
9 May 1926	Floyd Bennett (P); Richard E. Byrd (N)
11 May 1926	Roald Amundsen (expedition cmdr.); Lincoln Ellsworth (deputy); Umberto Nobile (captain); Hj. Riiser-Larsen (N), + 12
20 May 1927	Charles A. Lindbergh
4 June 1927	Clarence D. Chamberlin (P); Charles A. Levine (passenger-owner)
28 June 1927	Lester J. Maitland (P); Albert F. Hegenberger (N)
29 June 1927	Richard E. Byrd (N); Bert Acosta (P); Bernt Balchen (P); George O. Noville (R)
16 Apr. 1928	Carl Ben Eielson (P); George H. Wilkins (N)
31 May 1928	Charles E. Kingsford-Smith (P); Charles T. P. Ulm (P); Harry Lyon (N); James W. Warner (R)
11 Oct. 1928	Hugo Eckener + 38 crew, 19 passengers
1 Apr. 1930	W. H. Alexander (P); Lewis A. Yancey (N); Zeh Bouck (R)
1 Sep. 1930	Dieudonne Costes (P); Maurice Bellonte (N)
23 June 1931	Wiley Post (P); Harold Gatty (N)
20 May 1932	Amelia Earhart Putnam
9 July 1933	Charles Lindbergh (P, N); Anne Morrow Lindbergh (P, N, R)
3 Dec. 1934	C. T. P. Ulm (P); G. M. Littlejohn (P); J. L. Skilling (N, R)
30 Sep. 1935	? (P); P. V. H. Weems (N)
1 June 1937	Amelia Earhart Putnam (P); Frederick J. Noonan (N)
18 June 1937	Valery Chkalov (P); Alex Beliakov (N); George Baidukov (P, N, R)
10 July 1938	Howard Hughes (P); Thomas L. Thurlow (N); Harry P. M. Connor (N); + 2
6 Oct. 1938	Donald C. T. Bennett (P, N); Ian Harvey (P, R)

Aircraft	Route
4 Douglas single-engine biplanes; only 2 completed	Santa Monica–Seattle–west to Seattle
LZ 126/ZR III, airship	Lake Constance–Lakehurst, N.J.
Josephine Ford, Fokker trimotor	Spitsbergen Is.–North Pole–Spitsbergen Is.
Airship *Norge*	Spitsbergen Is.–North Pole–Pt. Barrow–Teller, Alaska
Spirit of St. Louis, Ryan single engine	New York–Paris
Columbia, Bellanca single engine	New York–Germany
Bird of Paradise, Fokker trimotor	San Francisco–Hawaii
America, Fokker trimotor	New York–Caen, France
Lockheed Vega single engine	Point Barrow–Spitsbergen
Southern Cross, Fokker trimotor	San Francisco–Hawaii–Suva–Brisbane
Graf Zeppelin	Lake Constance–Lakehurst–east around world to Lake Constance
Flying Laboratory, single engine, pontoons	New York–Bermuda
Point d'interrogation, Breguet biplane	Paris–New York
Winnie Mae, Lockheed Vega	New York–east to New York
Lockheed Vega	Teterboro, N.J.–Londonderry
Tingmissartoq, Lockheed Sirius single engine	New York–Greenland–Europe–Africa–Brazil–New York
Airspeed Envoy, twin engine	San Francisco– ?
O-2U	Maxwell Fld., Ala.–Langley Fld., Va.
Lockheed Electra twin engine	Miami–South America–Africa–New Guinea– ?
ANT-25 single engine	Moscow–Vancouver, Wash.
Lockheed 14	New York–east to New York
Mercury, Short seaplane, 4 engine	Dundee, Scotland–Alexander Bay, South Africa

NOTES

For complete bibliographic data on books, articles, and reports cited in the notes, refer to the bibliography, which is arranged alphabetically by author. If the author is unknown, the citation in the note is complete. In the bibliography, signed government reports appear under the author's name; unsigned government reports are alphabetized under the government and the principal office. Subsidiary offices are omitted in the notes.

CHAPTER 1: MARINE NAVIGATION BEFORE WORLD WAR I

1. See Harold Augustin Calahan, *The Sky and the Sailor;* Per Arne Collinder, *History of Marine Navigation;* Joseph Bushby Hewson, *A History of the Practice of Navigation;* and David W. Waters, *The Art of Navigation in England in Elizabethan and Early Stuart Times.*
2. For a more detailed description of the procedures and instruments treated in summary fashion in this chapter, see U.S., Navy Dept., *The American Practical Navigator;* or Great Britain, Admiralty, *Manual of Navigation, 1914.*
3. U.S., Navy Dept., *American Practical Navigator,* p. 16 for the quotation; Fritz E. Uttmark, "A New System of Navigation and Nautical Astronomy," p. 396.
4. "Modern Methods in Navigation and Nautical Astronomy," *Quarterly Review* 141 (Jan. 1876): 145; Albert B. Duval and Léo Hébrard, *Traité pratique de navigation aérienne,* p. 11; Fritz Gansberg, *Der Flugzeugkompass und seine Handhabung,* p. 33; U.S., War Dept., *Aerial Navigation,* p. 12.
5. "Sir William Thomson on Navigation," *Nature* [London] 15 (8 Mar. 1877): 403.
6. "Curves of Position for Determining the Place of a Ship at Sea," *Naval Science* 3 (1874): 39; "Sumner's Method at Sea," *Nature* [London] 14 (24 Aug. 1876): 346–48; H. A. Moriarty, "Navigation," pp. 271–72; Christopher Craddock, *Whispers from the Fleet,* p. 197; R. H. Curtiss, "The Rise of Navigation," p. 378.
7. U.S., Navy Dept., *American Practical Navigator,* pp. 115, 125–26; William Robert Martin, "Navigation," p. 296 for the quotation; Philip Van Horn Weems, *The Secant Time Sight,* p. 14.
8. W. R. Martin, "Navigation," p. 294 for the quotation; Theo Emil Sönnichsen, *Navigation und Seemannschaft im Seeflugzeug,* p. 12.

CHAPTER 2: THE NAVIGATION OF BALLOONS

1. The Société française de navigation aérienne published *Aéronaute* from 1864; the British Aeronautical Society its

Annual Report from 1866, its monthly *Aëronautical Journal* from 1897; the Deutscher Verein für Forderung der Luftschiffahrt (Berlin) from 1882, and the Flugtechnischer Verein (Vienna) from 1888, journals that became the monthly *Zeitschrift für Luftschiffahrt*.

2. Hermann W. L. Moedebeck, *Handbuch der Luftschiffahrt*, p. 63; Baden Fletcher Smyth Baden-Powell, *Ballooning as a Sport*, p. 113; A. de la Baume-Pluvinel, "La détermination du point en ballon; rapport sur l'état actuel de la question," p. 83.
3. André Beaumont, *My Three Big Flights*, p. 147; Hermann W. L. Moedebeck, "Aeronautische Landkarten," p. 299; Baden-Powell, *Ballooning as a Sport*, pp. 113–14 for the quotation; *Flight* 2 (18 June 1910): 465; 3 (22 July 1911): 641; 4 (24 Aug. 1912): 776.
4. Moedebeck, "Ueber das Landen mit Ballons," p. 273.
5. *Aeronautical World* 1 (1 Oct. 1902): 67, quoted in Walter W. Ristow, *Aviation Cartography; A Historico-Bibliographic Study of Aeronautical Charts*, pp. 4, 152; Charles-Édward Guillaume, "Rapport succinct sur les travaux scientifiques de la C.I.P.A.," pp. 63–64.
6. Moedebeck, "Aeronautische Landkarten." On the value of sounds for night ballooning, see also E. J. Saunière, "Ascension du ballon 'L'Alliance' le 25 janvier 1902," and A. Prevost, "Ascension du 'Touriste.'"
7. "La carte des canalisations électriques aériennes," *Aéronautique* 7 (Jan. 1908): 6.
8. Ristow, *Aviation Cartography*, p. 3.
9. Moedebeck, "Les cartes aéronautiques de la Fédération allemande," pp. 91–96; Moedebeck, "Die Karte des deutschen Luftschiffer-Verbandes," pp. 14–18.
10. *American Aeronaut* 1 (Aug. 1909): 33.
11. *Flight* 1 (5 June 1909): 336; Congrès international d'aéronautique, IVe, Nancy, septembre 1909, *Procès-verbaux; rapports & mémoires*, p. 469 for the quotation; "Commission de cartographie aéronautique du 4 septembre 1910," *Aérophile* 18 (1 Oct. 1910): 454; Ristow, *Aviation Cartography*, pp. 8, 80–81.
12. P. Pollacchi, "La carte aéronautique du Service géographique de l'armée"; *Flight* 3 (11 Feb. 1911): 123; "A Map for Military Aviators," ibid. (29 Apr. 1911): 373; "How to Find Your Way in the Air," *Scientific American* 105 (8 July 1911): 24–26; Bertram G. Cooper, "Aeronautical Maps and Signs"; "Aeronautic Maps Produced," *Aero; America's Aviation Weekly* 3 (28 Oct. 1911): 83, cited in Ristow, *Aviation Cartography*, p. 112.
13. Bertram G. Cooper, "Meeting of the International Commission for Aeronautical Maps, Brussels, 26 and 27 May, 1911"; *Scientific American* 105 (9 Sep. 1911): 226; Ristow, *Aviation Cartography*, pp. 5–6.
14. Hans Steffen and Otto Ber-

tram, *Orientierung auf Überlandflügen auf Grund praktischer Erfahrungen,* pp. 3–6; Pollacchi, "Carte aéronautique," p. 202.
15. Moedebeck, "Cartes aéronautiques," pp. 95–96; B. G. Cooper, "Meeting of the International Commission," p. 124; Cooper, "Aeronautical Maps and Signs," p. 143; F. W. Kaiser, "Navigation für Luftschiffe," p. 2; G. Kammerer, "La 'Photocarte' Scheimpflug," *Conquête de l'air* 9 (15 Aug. 1912): 7; Jean Buchot, "La carte aérienne," *Correspondent* 205 (25 Feb. 1913): 754–64; Kammerer and Buchot cited in Ristow, *Aviation Cartography,* pp. 8, 72.
16. Max Gasser, "Ueber die Luftschiffer-Kartenfrage," p. 11; Max Gasser, *Eine Flugkartenstudie,* charts bound in the back of the book; Pollacchi, "Carte aéronautique," p. 201; Beaumont, *My Three Big Flights,* p. 147; Karl Peucker, *Höhenschichtenkarten; Studien und Kritiken zur Lösung des Flugkartenproblems* (Stuttgart: Wittwer, 1910); Giovanni de Agostini, "La représentation du terrain dans les cartes aéronautiques," 5th International Congress of Aeronautics, Turin, 1911, *Procès-verbaux, rapports et mémoires* (Paris, 1912), pp. 375–82; Peucker and Agostini cited in Ristow, *Aviation Cartography,* pp. 8, 10.
17. Karl Bamler, "Deutsche Luftfahrerkarten," p. 60; Moedebeck, "Karte des deutschen Luftschiffer-Verbandes," p. 18.
18. Steffen and Bertram, *Orientierung auf Überlandflügen,* pp. 8-10.
19. Ristow, *Aviation Cartography,* p. 12 for the quotation; E. Lester Jones, "Aeronautic Maps."
20. A. Bestelmeyer, "Zur Benutzung des Kompasses im Ballon," p. 17; Ansbert Vorreiter and Hans Boykow, eds., *Volamekum; Handbuch für Luftfahrer,* pp. 71–72; Adolf Marcuse, *Astronomische Ortsbestimmung im Ballon,* pp. 13–14; Paul A. Meckel, "Ein Beitrag zur Frage der Balloninstrumente," p. 1151.
21. Léo Dex, "La détermination du point en ballon," p. 745; Bestelmeyer, "Benutzung des Kompasses," p. 17.
22. P. Marcillac, "Les appareils pour ascensions maritimes," p. 41; Dex, "Détermination du point," p. 744.
23. Dex, "Détermination du point," p. 744.
24. Ibid., p. 745.
25. J. Bourdin, "Sur un instrument analogue au compas aéronautique."
26. H. Bartsch v. Siegsfeld, "Astronomische Positionsbestimmungen im Frei-Ballon," pp. 2–5.
27. Bestelmeyer, "Benutzung des Kompasses"; Vorreiter and Boykow, *Volamekum,* pp. 73–74.
28. Thomas Baldwin, *Airopaidia; Containing the Narrative of a Balloon Excursion from Chester, the Eighth of September, 1785 . . . . To Which Is Subjoined, Mensuration of Heights by the Barometer, Made Plain . . .* (Chester: Printed for the author

by J. Fletcher, 1786), not seen.
29. A. H. Stuart, "An Alignment Chart for Obtaining Heights from Observations of Pressure and Temperature."
30. Henry Harrison Suplee, "Methods of Determining Aeroplane Altitudes," p. 330.
31. Napier John Gill, *The Flyer's Guide,* p. 5 for the quotation; "The Altimeter," *Aëronautical Journal* 8 (Apr. 1904): 42; *Flight* 5 (13 Sep. 1913): 1016.
32. Karl T. Fischer, "Ein neues Barometer ('Luftdruckaräometer')," p. 68; Émile Wenz, "Mesure des hauteurs atteintes par les aéroplanes," p. 11; L. Malevé, "Détermination de l'altitude en aviation"; Louis Paul Cailletet, "Appareil destiné à mesurer les hauteurs atteintes par les aérostats."
33. Marcillac, "Appareils pour ascensions maritimes," p. 41; Wilfrid de Fonvielle, "Nouvelle méthode altimétrique applicable en ballon"; Alphonse Berget, "Mesure des hautes altitudes"; Suplee, "Methods of determining aeroplane altitudes," p. 334; Bourdin, "Instrument analogue au compas aéronautique," p. 257; Bestelmeyer, "Benutzung des Kompasses," p. 18.
34. L. Favé, "Recherches sur les instruments et les méthodes propres à la détermination du point en ballon," p. 148, n. 1.
35. John M. Bacon, "Steering Balloons by Upper Air Currents," p. 173.
36. J. E. Capper, "The Gordon-Bennett Cup of 1906," pp. 16–18 for the quotations; C. V. Boys, "Position Finding without an Horizon," p. 111.
37. Dex, "Détermination du point," p. 746.
38. A. de la Baume-Pluvinel, "La détermination du point en ballon," *Aérophile* 12 (Jan. 1904): 20; Baldit, "Observations sur la méthode de détermination du point en ballon," p. 70; Herbert Nelson Eaton, "Aerial Navigation and Navigating Instruments," pp. 788–89; [Plath, C., firm, Hamburg], *C. Plath, 1862–1962,* p. 121; Favé, "Recherches sur les instruments," pp. 147–48.
39. Siegsfeld, "Astronomische Positionsbestimmungen," pp. 6–7 for the quotation; Dex, "Détermination du point," p. 748; la Baume-Pluvinel, "Détermination du point en ballon; rapport sur l'état actuel de la question," pp. 84, 87.
40. Dex, "Détermination du point," p. 747.
41. Edv. Jäderin, "Nivåsextant, konstruerad för Andrées Polarballong," for the translation of which I thank Mrs. Theodore Faurer, USAF Academy; Adolf Marcuse, "Die astronomische Ortsbestimmung im Ballon und ihre Bedeutung für die Luftschiffahrt," p. 21; Adolf Marcuse, *Astronomische Ortsbestimmung im Ballon,* p. 8; la Baume-Pluvinel, "Détermination du point en ballon; rapport sur l'état actuel de la question," p. 84; Alfred Wegener, "Astronomische Ortsbestimmung im Luftballon," pp. 116–17; Walter Leick, *Astronomische Ortsbestimmungen mit*

besonder Berücksichtigung der Luftschiffahrt, pp. 39–40.
42. A. de la Baume-Pluvinel, "Note sur l'emploi du sextant à niveau pour faire le point en ballon," pp. 64–65; Leick, *Astronomische Ortsbestimmungen*, pp. 40–41; W. Lindt, "Ein Libellenquadrant in neuer Form für astronomische Ortsbestimmungen"; E. Hartmann, "Ein Libellenquadrant in neuer Form für astronomische Ortsbestimmungen (Zahnbogen-Höhenmesser)."
43. F. J. B. Cordeiro, "Pulfrich Sextant"; [Plath, C.], *C. Plath, 1862–1962*, p. 119; Leick, *Astronomische Ortsbestimmungen*, pp. 41–42; Boys, "Position Finding," pp. 111–13; Favé, "Recherches sur les instruments," pp. 148–52.
44. Wegener, "Astronomische Ortsbestimmung," pp. 117–18; A. de la Baume-Pluvinel, "Recherches sur la détermination du 'Point,' et expériences de télégraphie sans fil en ballon," p. 504.
45. Walter Leick, "Praxis der Ortsbestimmung im Ballon," p. 995; Favé, "Recherches sur les instruments," pp. 145–46; Wegener, "Astronomische Ortsbestimmung," p. 117.
46. Leick, *Astronomische Ortsbestimmungen*, p. 84 for the quotation; Siegsfeld, "Astronomische Positionsbestimmungen," pp. 2–3; Guillaume, "Rapport succinct sur les travaux scientifiques," p. 64; la Baume-Pluvinel, "Détermination du point en ballon," *Aérophile* 12 (Jan. 1904): 20.
47. Siegsfeld, "Astronomische Positionsbestimmungen," p. 8.
48. Leick, *Astronomische Ortsbestimmungen*, p. 84; la Baume-Pluvinel, "Détermination du point en ballon," *Aérophile* 12 (Jan. 1904): 21.
49. Favé, "Recherches sur les instruments," pp. 159–61; Leick, *Astronomische Ortsbestimmungen*, p. 85.
50. Favé, "Recherches sur les instruments," p. 162 for the quotation; Guillaume, "Rapport succinct sur les travaux scientifiques," p. 67.
51. Such a star graph, computed for Paris, is illustrated in Favé, "Recherches sur les instruments," p. 170.
52. Such a graph, at too small a scale for practical use, is illustrated ibid., p. 173.
53. The charts are illustrated ibid., opposite p. 162.
54. Leick, *Astronomische Ortsbestimmungen*, pp. 85–87, 126; Alfred Brill, "Ein Verfahren zur Auswertung astronomischer Positionsbestimmungen"; Alfred Brill, "Über eine neue einheitliche Methode zur nautischen und aeronautischen Ortsbestimmung aus Gestirnsmessungen"; Boys, "Position Finding without an Horizon," pp. 231–32; Vorreiter and Boykow, *Volamekum*, pp. 55–57; Oskar Voigt, "Zur astronomische Ortsbestimmung"; Oskar Voigt, *Ein Beitrag zur astronomischen Ortsbestimmung im Luftfahrzeuge*.
55. La Baume-Pluvinel, "Note sur l'emploi du sextant," p. 66; Baldit, "Observations sur la méthode de détermination du point," pp. 67–68.

56. Eugen Alt, "Astronomisches Ortsbestimmung im Luftschiff."
57. Marcuse, *Astronomische Ortsbestimmung,* pp. 18-19; Leick, *Astronomische Ortsbestimmungen,* pp. 94-95.
58. Johannes Möller, "Die astronomische Ortsbestimmung im Luftschiff"; P. B. Fischer, "Astronomische Ortsbestimmung im Ballon nach Schwarzschild"; K. Schwarzschild, "Tafeln zur astronomischen Ortsbestimmung, von A. Kohlschütter"; Walter D. Robinson, Review of Émile Guyou, *Nouvelles tables de navigation;* R. Süring, ed., *Moedebecks Taschenbuch zum praktischen Gebrauch für Flugtechniker und Luftschiffer,* p. 569 for the quotation.
59. Schwarzschild, "Tafeln zur astronomischen Ortsbestimmung"; Philip Van Horn Weems to P. F. Everitt, 5 May 1931, Weems residence, Annapolis (hereafter cited as WRes).
60. H. H. Kritzinger, "Die Uhr als Orientierungsmittel im Ballon"; *Flight* 5 (18 Oct. 1913): 1145.
61. Arendt, "Ueber die Bedeutung magnetischer Beobachtungen im Ballon," p. 209; Bidlingmaier, "Über die magnetische Ortsbestimmung im Ballon"; Marcuse, *Astronomische Ortsbestimmungen,* p. 6.
62. Möller, "Astronomische Ortsbestimmung," p. 971; la Baume-Pluvinel, "Détermination du point en ballon; rapport sur l'état actuel," p. 85.
63. Ristow, *Aviation Cartography,* pp. 74, 113; Gustav W. Hamel and Charles Cyril Turner, *Flying; Some Practical Experiences,* p. 80; *Flight* 11 (29 May 1919): 720.
64. Abbott Lawrence Rotch and Andrew Henry Palmer, *Charts of the Atmosphere for Aeronauts and Aviators;* Eaton, "Aerial Navigation and Navigating Instruments," p. 770 for the quotation.

CHAPTER 3: NAVIGATION OF AIRSHIPS AND AIRPLANES BEFORE WORLD WAR I

1. J. Voyer, "Note sur la vitesse des navires aériens par rapport au sol," pp. 59, 66; Paul Delens, "La dérive dans le vent"; Delens, "Étude des parcours aériens."
2. Louis Toft, "Effect of the Wind on Flight Speeds"; "Sailor," "La boussole et les cartes en aéronavigation," p. 326; G. Daloz, "Boussole aérienne," *Aérophile* 18 (1 Dec. 1910): 533; Daloz, "La boussole aérienne," *Revue aérienne* 3 (25 Nov. 1910): 652-58.
3. "Sailor," "Boussole et les cartes," p. 326.
4. Hamel and Turner, *Flying,* pp. 78, 85; Gill, *Flyer's Guide,* p. 23.
5. *Daily Mail,* 26 July 1909, quoted in Neville Duke and Edward Lanchbery, eds., *The Saga of Flight; an Anthology,* p. 95; *Flight* 1 (31 July 1909): 458.
6. M. Tevis, "Return Flight of Carrier Pigeons"; A. Jurmont, "L'instinct de l'orientation chez les aviateurs."

7. William James Vivian Branch and E. Brook-Williams, *A Short History of Navigation,* p. 11. This book gives no sources. However, Cody carried one Cmdr. Osborne, R.N., as a passenger on a flight in Sep. 1909 (*Flight* 1 [2 Oct. 1909]: 615). Subsequently a Capt. Creagh-Osborne, R.N., is identified as a pioneer in aero compass experimental work and Superintendent of the Admiralty's Compass Department (*Flight* 4 [5 Oct. 1912]: 891, and 6 [26 June 1914]: 689). If Creagh-Osborne was Cody's passenger, they probably experimented with a marine compass.
8. "Les instruments de précision à l'exposition d'aéronautique de 1909," *Aérophile* 18 (1 Jan. 1910): 15; *Flight* 3 (25 Mar. 1911): 261–62; *Scientific American* 104 (4 Mar. 1911): 230.
9. Arthur J. Hughes, *History of Air Navigation,* p. 100. This book lacks the scholarly paraphernalia, but Mr. Hughes, of Henry Hughes & Son, Ltd., was long closely associated with the manufacture of aeronautical instruments. The firm's founder supplied watches to Capt. Bligh. See also Frank Creagh-Osborne, *The Magnetic Compass in Aircraft,* pp. 26–27.
10. *Flight* 3 (28 Jan. 1911): 78, reprinted from *Naval and Military Record;* Eric Hollocoombe Clift, "The Magnetic Compass; Its Construction and Use," p. 164; see also "Sailor," "Boussole et les cartes."
11. C. O. [Creagh-Osborne ?], Letter to the editor, *Flight* 3 (6 May 1911): 409; Clift, "Magnetic Compass," p. 165.
12. Clift, "Magnetic Compass," p. 164 for the quotation; *Scientific American* 104 (4 Mar. 1911): 230; William Armstrong, *Pioneer Pilot,* p. 48.
13. Eric Hollocoombe Clift, Letter to the editor, *Flight* 3 (13 May 1911): 429 for the quotation; ibid. (1 Apr. 1911): 293; "A New Aeroplane Compass," *Aëronautical Journal* 15 (Apr. 1911): 91; Gansberg, *Flugzeugkompass und seine Handhabung,* pp. 15–16.
14. Hughes, *History of Air Navigation,* p. 100; Gordon M. B. Dobson, "Design of Instruments for Navigation of Aircraft," p. 372; Creagh-Osborne, *Magnetic Compass in Aircraft,* p. 29.
15. Robert Eyb, *Fliegerhandbuch,* pp. 258–59; Gansberg, *Flugzeugkompass und seine Handhabung,* pp. 20–24; Gustav W. Hamel, Letter to the editor, *Flight* 3 (13 May 1911): 429; ibid. (12 Aug. 1911): 699; 7 (13 Aug. 1915): 591–92; Henri Salmet, "My Paris Flight"; "How to Find Your Way in the Air," *Scientific American* 105 (8 July 1911): 26; Hans Boykow, "Orientierung und Navigation im Luftfahrzeug," p. 532; v. Renesee, "Gedanken über die Ausbildung eines Beobachtungs-Offiziers," p. 88; Samuel S. Pierce, "The Use of the Compass in Aviation," p. 111; Gill, *Flyer's Guide,* p. 17.
16. Geoffrey de Havilland, *Sky Fever,* p. 85; Beaumont, *My Three Big Flights,* p. 15; quoted

in Harry Egerton Wimperis, *A Primer of Air Navigation*, p. 28.
17. B. C. Hucks, "A Further Three Years' Flying Experience," p. 590.
18. "Annual Report of the Advisory Committee for Aeronautics, 1915–16," *Flight* 8 (24 Aug. 1916): 720–22.
19. Kaiser, "Navigation für Luftschiffe," p. 3.
20. Hermann W. L. Moedebeck, *Pocket-Book of Aeronautics*, p. 360.
21. Bertram Dickson, "A Safety Speed-Alarm for Aeroplanes."
22. Winslow H. Herschel, "The Pitot Tube and Other Anemometers for Aeroplanes," p. 94.
23. "A Self-Timing Anemometer," *Flight* 3 (6 May 1911): 402, describes an instrument that was only semiautomatic.
24. Wilke, "Speed Measuring Instruments for Aeroplanes"; Franklin L. Hunt, "Aircraft Speed Instruments: Airspeed Indicators," p. 565; *Flight* 6 (4 Apr. 1914): 360.
25. *Flight* 5 (1 Mar. 1913): 256; Willis Stetson Fitch, *Wings in the Night*, p. 100; Herschel, "Pitot Tube and Other Anemometers," p. 99; Hunt, "Aircraft Speed Instruments," pp. 575–77.
26. Such an instrument is illustrated in *Flight* 4 (14 Sep. 1912): 840; see also Horace Darwin, "Scientific Instruments, Their Design and Use in Aeronautics," p. 599; Herschel, "Pitot Tube and Other Anemometers," pp. 79–84, 99–100.
27. A. P. Thurston, "The Measurement of Air Speed," p. 498 for the quotation; Horace Darwin, "Scientific Instruments Used in Aeronautics"; Bertram G. Cooper, "Air-Speed Indicators"; Harris Booth, "The Theory of the Gravity-Controlled Air Speed Indicator as Applied to Dynamic Lift Flying Machines"; Editorial, *Flight* 6 (3 Jan. 1914): 1; Harris Booth, "A Warning to Pilots"; Leonard Bairstow, "Discussion of the Action of Different Types of Air-Speed Indicators."
28. Thurston, "Measurement of Air Speed," p. 521; Hamel and Turner, *Flying*, p. 127; U.S. National Advisory Committee for Aeronautics, *1st Annual Report, 1915*, p. 14 for the quotation.
29. L. Graham Davies, "Steering by Compass"; Moedebeck, *Pocket-Book of Aeronautics*, p. 359.
30. Daloz, "Boussole aérienne," *Aérophile* 18 (1 Dec. 1910): 531–33; René Ozouf, "The Daloz Compass"; *Scientific American* 105 (8 July 1911): 26.
31. Steffen and Bertram, *Orientierung auf Überlandflügen*, pp. 46–48; Darwin, "Scientific Instruments, Their Design and Use," p. 600 for the quotation; *Flight* 4 (20 July 1912): 661; 5 (13 Sep. 1913): 1016.
32. *Flight* 3 (1 Apr. 1911): 280, (22 July 1911): 641; "Miss Ruth Law Now Holds American Non-Stop Cross-Country

Record," *Aerial Age Weekly* 4 (27 Nov. 1916): 277.

33. Paul Bewsher, *"Green Balls"; The Adventures of a Night-Bomber,* p. 116; Hermann Koehl, James C. Fitzmaurice, and Guenther von Huenefeld, *The Three Musketeers of the Air,* p. 139; Aeronautic Library advertisement, *Aerial Age Weekly* 5 (23 Apr. 1917): 206; Steffen and Bertram, *Orientierung auf Überlandflügen,* p. 13; Vorreiter and Boykow, *Volamekum,* pp. 78–79; Maurice Percheron, *Utilisation de la carte et de la boussole en aviation,* p. 9.

34. Beaumont, *My Three Big Flights,* p. 148; Steffen and Bertram, *Orientierung auf Überlandflügen,* p. 17; Hamel and Turner, *Flying,* p. 89; Eyb, *Fliegerhandbuch,* p. 263 for the quotation.

35. Steffen and Bertram, *Orientierung auf Überlandflügen,* p. 23.

36. Hamel and Turner, *Flying,* p. 88; "Sperry Charts for Aviators," *Flight* 9 (6 Sep. 1917): 928; Percheron, *Utilisation de la carte,* pp. 7–9 for the quotation.

37. Walter Alexander Raleigh, *The War in the Air,* 1: 193, 196 for the quotations; Steffen and Bertram, *Orientierung auf Überlandflügen,* pp. 1–2, 23–29.

38. Beaumont, *My Three Big Flights,* p. 91.

39. Steffen and Bertram, *Orientierung auf Überlandflügen,* pp. 43–45; Salmet, "My Paris Flight"; Percheron, *Utilisation de la carte,* pp. 26–27.

40. "Mr. Cody's Flight to Manchester," *Flight* 1 (9 Oct. 1909): 627; ibid. 2 (7 May 1910): 351 for the quotation.

41. Beaumont, *My Three Big Flights,* pp. 77, 146.

42. *Flight* 2 (1 Jan. 1910): 12, (29 Oct. 1910): 892, (31 Dec. 1910): 1081; "How to Find Your Way in the Air," *Scientific American* 105 (8 July 1911): 25; Frankenberg, "Kirchturmuhren, ein Orientierungsmittel für Luftfahrzeuge."

43. Georg Rothgiesser, "Wegweiser für Luftschiffer"; Frankenberg, "Terrestrisches Karten-Orientierungs-System"; Paul A. Meckel, "Orientierungssystem"; F. Rasch, "Luftfahrerkarte und Orientierungszeichen"; *Scientific American Supplement* 76 (20 Sep. 1913): 179; "Sailor," "Boussole et les cartes," p. 327 for the quotation.

44. Boykow, "Orientierung und Navigation," p. 530; Editorial, *Flight* 5 (6 Dec. 1913): 1316.

45. George O. Squier, "Present Status of Military Aeronautics"; "Military Aeronautics in Germany," *Flight* 6 (2 Oct. 1914): 997-1002; ibid. 2 (24 Dec. 1910): 1059; 3 (9 Dec. 1911): 1068; 5 (18 Oct. 1913): 1154; Raoul Volens, "Phares pour la navigation aérienne," pp. 250–51; Alfred Gradenwitz, "Lighthouses for the Aerial Navigator; Guiding the Airman at Night."

46. *Scientific American* 110 (7 Mar. 1914): 218–19.

47. Steffen and Bertram, *Orientierung auf Überlandflügen,* pp.

1–2; Hamel and Turner, *Flying*, p. 98.
48. Davies, "Steering by Compass," p. 775.
49. "The first experiment in scientific aerial navigation was carried out from Eastbourne aerodrome, when the proprietor, Mr. Fowler, took Mr. Rainey, one of the watch-keeping officers of the Royal Mail Steam Packet Company, for a flight out to sea. Mr. Rainey used a sextant and a chronometer, and determined his latitude by the double altitude method. He claimed to be able to fix the machine's position to a quarter of a mile, whilst out of sight of land." R. Dallas Brett, *History of British Aviation, 1908–1914*, p. 203. This book was written from the office files of *Flight* magazine. The incident is not dated, but previous and succeeding events place it at about April 1913. Given the details supplied by Brett, the flight probably took place, but I cannot accept the accuracy claimed.
50. Hamel and Turner, *Flying*, p. 88.
51. Ibid., pp. 98–100; Steffen and Bertram, *Orientierung auf Überlandflügen*, pp. 23, 25; Renesse, "Gedanken über die Ausbildung," p. 88 for the quotation.
52. Steffen and Bertram, *Orientierung auf Überlandflügen*, pp. 14, 24; W. S. Brachner, "Training in Military Aviation."
53. Walter Wellman, *The Aerial Age*, pp. 174, 184–85, 220–22, 260, 276.
54. Ibid., pp. 262–69, 335–67, quotations from pp. 350, 275; Eaton, "Aerial Navigation and Navigating Instruments," pp. 763, 788.
55. James V. Martin, "Across the Atlantic by Aeroplane"; Hans Boykow, "Transatlantic Flight."
56. *Flight* 6 (26 June 1914): 689–90; *Scientific American* 111 (25 July 1914): 58–59.

CHAPTER 4: NAVIGATION IN WORLD WAR I

1. Raleigh, *War in the Air*, 1: 7.
2. Ibid., pp. 298–300.
3. *Flight* 10 (7 Feb. 1918): 142 for the quotation; ibid. 11 (10 July 1919): 898; C. J. Stewart, "Instruments," p. 209; Duval and Hébrard, *Traité pratique*, p. E.
4. "The Military Aeroplane," a discussion organized by the Aeronautical Society and the Royal United Service Institution, 6 Dec. 1911, *Flight* 3 (16 Dec. 1911): 1085-87; Raleigh, *War in the Air*, 1: 260.
5. Henry Albert Jones, *The War in the Air*, 3: 292, 294; Maj. Gen. W. L. Kenly, "Report of Year Ending 30 June 1918"; Brachner, "Training in Military Aviation," p. 242.
6. Raleigh, *War in the Air*, 1: 443; H. A. Jones, *War in the Air*, 3: 298–99; Georg Paul Neumann, ed., *Die deutschen Luftstreitkräfte im Weltkriege*, p. 193; Georg Wilhelm Haupt-Heydemarck, *Double-Decker C.666*, p. 14; Koehl, Fitzmaurice, and von Huenefeld, *Three Musketeers*, pp. 11–13.
7. Elisha Noel Fales, *Learning to Fly in the U. S. Army*, pp. 92–93 for the quotation; James

Dundas White, *Steering by the Stars for Night-Flying, Night-Marching and Night Boat-Work*, pp. 4, 8; Charles Beverley Benson, *Map Reading for Aviators*, pp. 32–33; Samuel Edward Gideon, *Map Reading*, p. 25; Sandham Graves, *The Lost Diary*, pp. 73–74; Robert E. Hartz, "Map Reading by Aviators"; Creagh-Osborne, *Magnetic Compass in Aircraft*, p. 15.

8. U.S., War Dept., *Aerial Navigation*, p. 5; Creagh-Osborne, *Magnetic Compass in Aircraft*, p. 10; *Aerial Age Weekly* 5 (2 July 1917): 520; Report of an investigation of the Royal Flying Corps, 1916, quoted in Henry Woodhouse, *Textbook of Naval Aeronautics*, p. 129; "The Lost Night Bomber," *Flight* 11 (27 Feb. 1919): 290.

9. Charles Frederick Snowden Gamble, *The Story of a North Sea Air Station*, pp. 163, 165; H. A. Jones, *War in the Air*, 3: 170–71.

10. Duval and Hébrard, *Traité pratique*, p. 49; Mayo D. Hersey, "General Classification of Instruments and Problems Including Bibliography," p. 483.

11. Steffen and Bertram, *Orientierung auf Überlandflügen*, p. 43; Albert Edward Dixie, *Air Navigation for Flight Officers*, pp. 168–69.

12. Duval and Hébrard, *Traité pratique*, pp. 40–42. Albert B. Duval, "Le cercle calculateur de routes et de vitesses," describes a different instrument for the same purpose—an analog computer that constructed a parallelogram of velocities, with track and groundspeed being along the diagonal from the origin.

13. U.S., War Dept., *Aerial Navigation*, pp. 27–29; Stanton Freeland Card, *Air Navigation, Notes and Examples*, p. 58; Hunt, "Aircraft Speed Instruments," p. 580; Harry Egerton Wimperis, "Navigational Equipment for Long-Distance Flights," p. 759; Great Britain, Air Ministry, *Air Navigation Instruments (Excluding Compasses)*, pp. 10–15.

14. Duval and Hébrard, *Traité pratique*, pp. 33–34; Eaton, "Aerial Navigation and Navigating Instruments," pp. 780–81; Great Britain, Air Ministry, *Air Navigation Instruments*, pp. 18–19.

15. Horace Darwin, "The Static Head Turn Indicator for Aeroplanes."

16. Keith Lucas, "Report on the Errors of Compasses on Aeroplanes"; "Annual Report of the Advisory Committee for Aeronautics, 1915–16," *Flight* 8 (24 Aug. 1916): 720–22; S. G. Starling, "The Equilibrium of the Magnetic Compass in Aeroplanes"; R. L. Sanford, "Direction Instruments; the Testing and Use of Magnetic Compasses for Airplanes," p. 615.

17. John A. C. Warner, "Direction Instruments; Aircraft Compasses—Description and Classification," pp. 620–21, 624–26; Hughes, *History of Air Navigation*, pp. 100–101.

18. Warner, "Direction Instruments," p. 627; Hughes, *His-*

tory of Air Navigation, p. 101.
19. *Scientific American Monthly* 2 (Nov. 1920): 274; Walter Friedensburg, "The Distant Compass"; Warner, "Direction Instruments," pp. 637–40.
20. Darwin, "Static Head Turn Indicator," pp. 1429–30; Harry Egerton Wimperis, "Air Navigation," p. 44; George William Haddow and Peter Michael Grosz, *The German Giants; the Story of the R-Planes*, pp. 27, 231–32; Duval and Hébrard, *Traité pratique*, pp. 36–39.
21. U.S., National Advisory Committee for Aeronautics, *3d Annual Report, 1917*, pp. 25–26; Albert Francis Zahm, "Development of Air Speed Nozzles"; Great Britain, Ministry of Munitions, "Report on the Friedrichshafen Bomber," p. 796; U.S., War Dept., *Air-Speed Meters for Aerial Navigation*, pp. 7–9; Mayo D. Hersey, F. L. Hunt, and H. N. Eaton, "The Altitude Effect on Air Speed Indicators," pp. 693–94; Hunt, "Aircraft Speed Instruments," pp. 563–71.
22. U.S., Army, *Technical Order #1* (Dayton: Oct. 1918): 65 and Duval and Hébrard, *Traité pratique*, p. 35, for the quotations; also "Pioneer Aircraft Instruments," *Aerial Age Weekly* 11 (3 May 1920): 257.
23. *Flight* 7 (5 Feb., 12 Mar. 1915): 102, 186 for the quotations; P. R. Jameson, "The Development of the Tycos Altimeter"; A. H. Mears, H. B. Hendrickson, and W. G. Brombacher, "Altitude Instruments," pp. 502–05; Report of an investigation of the Royal Flying Corps, 1916, quoted in Woodhouse, *Textbook of Naval Aeronautics*, p. 130.
24. *Flight* 2 (3 Sep. 1910): 719; E. Jones, Letter to the editor, ibid. (17 Sep. 1910): 762; Claudio Piumatti, "Aeroplanes in Naval Warfare"; "Bomb-Dropping from Aeroplanes," *Flight* 6 (18 Dec. 1914): 1213–14; Riley E. Scott, "Dropping Bombs from Flying Machines"; *Scientific American* 107 (14 Sep. 1912): 214; Alexander Büttner, "Throwing Bombs from Airships."
25. Harry Egerton Wimperis, "Air Navigation; the Most Important of the Unsolved Problems Relating to Aviation," p. 482, implies that such attacks were "exceptional," but he must have been unduly influenced by his own driftmeter-bombsight, which made attacks from all directions possible. The instrument became available only toward the end of the war. The body of World War I bombing literature stresses attacking with the wind, or directly against it, for more accurate results, in spite of the advantage such restricted tactics gave the defense.
26. "Scientific Bomb Dropping," *Scientific American Supplement* 81 (22 Apr. 1916): 260, reprinted from *The Illustrated War News*; "Bomb Throwing from Aircraft," *Aerial Age Weekly* 4 (9 Oct. 1916): 98, 106, trans. from *Technische Rundschau*; H. A. Jones, *War in the Air*, 2: 119–20; Neumann, *Deutschen Luftstreit-*

27. Fitch, *Wings in the Night*, pp. 160–61; Bewsher, *"Green Balls,"* pp. 76, 129–30; Jean-Abel Lefrance, "Dropping Bombs from Aeroplanes"; Lefrance, "The German Goerz Bombing Sight"; "Air Raid over Zeebrugge; an Observer's Own Record," *Flight* 10 (18 July 1918): 791–92; Haddow and Grosz, *German Giants*, p. 93.
28. E. J. Loring, "Bombing and Bombing Sights," p. 49 for the quotation; Lefrance, "Dropping Bombs from Aeroplanes," p. 108; Carl Dienstbach, "The Gyrotelescope."
29. Eaton, "Aerial Navigation and Navigating Instruments," p. 774.
30. "The 'Crocco' Route Indicator and Its Noteworthy Applications," *Aerial Age Weekly* 7 (18 Mar. 1918): 54–56, trans. from *Rivista d'aeronautica*; Eaton, "Aerial Navigation and Navigating Instruments," pp. 773–76; Wimperis, *Primer of Air Navigation*, pp. 41–42.
31. Eaton, "Aerial Navigation and Navigating Instruments," pp. 771, 777; Elmer Ambrose Sperry, "Aerial Navigation over Water," chap. 16 of Woodhouse, *Textbook of Naval Aeronautics*, p. 108 for the quotation; Neil MacCoull, "The Sperry Drift Indicator."
32. *Scientific American Supplement* 66 (26 Dec. 1908): 406; *Flight* 2 (1, 8 Oct., 10 Dec. 1910): 802, 824, 1017; 3 (5 Aug., 30 Dec. 1911): 685, 1144; *Air Service Weekly News Letter* 1 (14 Dec. 1918): 4.
33. La Baume-Pluvinel, "Recherches sur la détermination du 'Point,' " p. 504 for the quotation; Fritz Lux, "Wellentelegraphische Ortsbestimmung für die Luftschiffahrt"; Paul Ludewig, *Die drahtlose Telegraphie im Dienste der Luftfahrt*, p. 59.
34. *Flight* 5 (2 Aug. 1913): 844; "Wireless Telegraphy and Aircraft," ibid. 6 (13 Nov. 1914): 1119–21; W. Reginald Dainty, "Radio-Telegraphic Apparatus for Aircraft."
35. T. Thorne Baker, "Wireless Telegraphy and Flight"; "Models," a column, *Flight* 6 (17 Jan. 1914): 76; Marcus Dyce Manton, "Wireless Equipment on Aircraft"; Raleigh, *War in the Air*, 1: 229.
36. William Dubilier, "Wireless Communication and Aeronautics"; Frederick Hugh Sykes, *From Many Angles*, p. 139; Gamble, *Story of a North Sea Air Station*, p. 44; H. A. Jones, *War in the Air*, 2: 82.
37. Louis Heathcote Walter, *Directive Wireless Telegraphy*, pp. 2, 22–23; J. Erskine-Murray, "Determining Position at Sea by Wireless."
38. "Wireless Direction Finder," *Electrical World* 64 (1 Aug. 1914): 245–46; Walter, *Directive Wireless Telegraphy*, pp. 11–14, 75-77.
39. Ludewig, *Drahtlose Telegraphie*, pp. 59–62; Sidney Herbert Long, *Navigational Wireless*, pp. 51–53.
40. Erskine-Murray, "Determining Position at Sea," pp. 950–

51; Raleigh, *War in the Air*, 1: 363; Robert G. Skerrett, "The Radio Compass and Navigation," p. 89.
41. Neumann, *Deutschen Luftstreitkräfte*, pp. 193–94; H. A. Jones, *War in the Air*, 2: 83. The British Sterling set was short range, but adequate for artillery spotting. The weight of radio equipment for use on long-range airplanes continued to be significant throughout the war.
42. Jerome Lachenbruch, "The Wireless Compass"; Skerrett, "Radio Compass," p. 89; H. A. Jones, *War in the Air*, 2: 391–92; 3: 129, 187, 319; 4: 115–16; Douglas Hill Robinson, *The Zeppelin in Combat*, p. 72.
43. Lord Weir of Eastwood, "Some Developments in Aircraft Design and Application during the War," p. 1052; D. H. Robinson, *Zeppelin in Combat*, p. 211 for the quotation.
44. Treusch von Buttlar-Brandenfels, Horst J. L. O. *Zeppelins over England*, p. 201; Walter, *Directive Wireless Telegraphy*, p. 46; D. H. Robinson, *Zeppelin in Combat*, p. 298.
45. James Robinson, "Directional Wireless in Air Navigation."
46. Eaton, "Aerial Navigation and Navigating Instruments," pp. 799-800; Walter, *Directive Wireless Telegraphy*, pp. 66–67, 99–102.
47. E. Bellini, "Errors of Direction Finders," p. 451; Frank Adcock, "Some Early Observations on Aircraft with the Four-Aerial Direction-Finder"; J. Robinson, "Directional Wireless in Air Navigation," p. 963; Eaton, "Aerial Navigation and Navigating Instruments," pp. 763–64.
48. D. H. Robinson, *Zeppelin in Combat*, p. 75. This book, based on documents from the German naval archives captured by the Allies in 1945, is rich in technical detail, and I have used it extensively.
49. Treusch von Buttlar, *Zeppelins over England*, p. 63; Neumann, *Deutschen Luftstreitkräfte*, p. 376; D. H. Robinson, *Zeppelin in Combat*, Appendix A.
50. D. H. Robinson, *Zeppelin in Combat*, pp. 59–63; H. A. Jones, *War in the Air*, 3: 90–91.
51. H. A. Jones, *War in the Air*, compares the detailed British records of the Zeppelin raids with the German account, sometimes taken from the German official history, sometimes from German crew reports. Jones devotes two volumes to maps showing the track of each raider, based on British records; even the impact of all identifiable bombs is plotted.
52. D. H. Robinson, *Zeppelin in Combat*, pp. 72, 79–80.
53. Ibid., pp. 4, 74; H. A. Jones, *War in the Air*, 3: 83–84, 144–45; [Heinrich Bahn], "In a German Airship over England," p. 108.
54. H. A. Jones, *War in the Air*, 3: 210–13; D. H. Robinson, *Zeppelin in Combat*, pp. 141–43.
55. D. H. Robinson, *Zeppelin in Combat*, p. 230.
56. Ibid., pp. 265–80; H. A. Jones, *War in the Air*, 5: 92–102; Treusch von Buttlar, *Zeppelins*

over England, pp. 155–60; Waldemar Kölle, "Marineluftschiffe im Kriege, in Sturm und Not," pp. 53–60.
57. Treusch von Buttlar, Zeppelins over England, p. 184 for the quotation; D. H. Robinson, Zeppelin in Combat, pp. 71, 193, 285, 291, 356; A. Wedemeyer, "Astronomische Ortsbestimmung," in Reinhard Süring and K. Wegener, eds., Moedebecks Taschenbuch für Flugtechniker und Luftschiffer, p. 344; Haddow and Grosz, German Giants, p. 231.
58. D. H. Robinson, Zeppelin in Combat, pp. 298, 333–34; Treusch von Buttlar, Zeppelins over England, p. 201; H. A. Jones, War in the Air, 5: 133.
59. Léo Hébrard, "Le bombardement de nuit et les problèmes de la navigation aérienne," p. 159; Maj. Evelyn B. Gordon, "Some Notes on Bombing Attacks," Hq., RFC, 21 Dec. 1915, in H. A. Jones, War in the Air, 2: 462–63; ibid., 5: 28.
60. A. R. Kingsford, Night Raiders of the Air, p. 98 and, for the quotation, p. 127; Bewsher, "Green Balls," pp. 65, 73, 128.
61. Robert H. Reece, Night Bombing with the Bedouins, pp. 46–48, 88–89; Kingsford, Night Raiders, pp. 97–99; Fitch, Wings in the Night, pp. 135–36; Flight 10 (27 June 1918): 720; Bewsher, "Green Balls," p. 53 for the quotation.
62. Bewsher, "Green Balls," pp. 241–48; Haddow and Grosz, German Giants, p. 18; Hébrard, "Bombardement de nuit," p. 161.

63. Hébrard, "Bombardement de nuit," p. 161; Duval, "Cercle calculateur," p. 157.
64. L. Dunoyer, "Le point estimé en navigation aérienne," pp. 144–45; Reece, Night Bombing, pp. 57–60.
65. Hébrard, "Bombardement de nuit," pp. 162–63; Hugh M. Trenchard, "The Work of the I. A. F.," p. 53; Reece, Night Bombing, pp. 76–77.
66. Raymond H. Fredette, The Sky on Fire, pp. 124, 134–35, 193; Haddow and Grosz, German Giants, pp. 34–35, 187, 231–32.
67. Night bomber crews have difficulty analyzing the results of their own work, particularly when bombing cities. But for an account by a German crew that made a gliding attack in daylight and completely missed their target because of a wind shift at the lower level, see Haupt-Heydemarck, Double-Decker C.666, p. 57.
68. Arthur Halsted, "Italian Semi-Rigid Airships Observed at the Front," p. 2; H. A. Jones, War in the Air, 6: 168, citing reports of the naval commander, Dunkirk, to the Air Board, Aug. 1917, and Trenchard's written comments on those reports, Sep. 1917; Trenchard, "Work of the I. A. F.," p. 53.
69. Charles Webster and Noble Frankland, The Strategic Air Offensive against Germany, 1: 227–28; 4: 205; Plans note, 16 Apr. 1924, ibid., 1: 48 for the quotation.
70. H. A. Jones, War in the Air, 4: 18–19; Gamble, Story of a North Sea Air Station, p. 354.

71. P. I. X. [T. D. Hallam], *The Spider Web,* pp. 26–32; H. A. Jones, *War in the Air,* 4: 53–54; U.S., Navy Dept., *Pocket Manual for Seaplane Pilots,* p. 17; Bewsher, "Green Balls," pp. 7–8, 11; Sönnichsen, *Navigation und Seemannschaft im Seeflugzeug,* pp. 69–70; Dixie, *Air Navigation for Flight Officers,* pp. 177–79.
72. Paul J. Haaren, "The Seagull Flies," p. 489; P. I. X., *Spider Web,* pp. 44, 117–18.
73. Gamble, *Story of a North Sea Air Station,* p. 34; P. I. X., *Spider Web,* pp. 72–73; Wimperis, "Air Navigation," pp. 42–43.
74. Duval and Hébrard, *Traité pratique,* p. 2.
75. *Aeronautics* 16 (8 May 1919): 482–87; see also Great Britain, Committee on Education and Research in Aeronautics, Report dated 12 Dec. 1919, p. 8.
76. H. A. Jones, *War in the Air,* 5: 429, 447; 6: 173; Editorial, *Flight* 11 (10 July 1919): 898.

CHAPTER 5: OVERLAND AIRLINES AND RADIO NAVIGATION

1. "To Establish Four Airways across the Continent and Use Mail Carrying Planes to Keep Manufacturers Busy and Train Aviators," *Aerial Age Weekly* 6 (19 Nov. 1917): 420; Frederick Hugh Sykes, "Some Empire Aspects of Aviation"; Sykes, *From Many Angles,* p. 281, citing his memo as first Controller-General of Civil Aviation to the Air Ministry, July 1919.
2. "Lord Weir on the Future of Flying," abstract of speech at Manchester, 20 Dec. 1918, *Flight* 11 (2 Jan. 1919): 16.
3. Ross Smith, *14,000 Miles through the Air,* p. 93 for the quotation, also pp. 12, 23–26, 130–31; Hughes, *History of Air Navigation,* pp. 46–47.
4. *Flight* 14 (13, 20 Apr. 1922): 216, 232.
5. Frida Helena Brackley, comp., *Brackles,* pp. 39, 41, 46; "Some Notes on the Transatlantic Voyage of R 34," *Flight* 11 (3 July 1919): 890; Interview of Maj. Scott, ibid. (17 July 1919): 950.
6. Henry Norris Russell, "On the Navigation of Airplanes," pp. 129, 148.
7. Francisco Radler de Aquino, Lt., Brazilian Navy, *Altitude and Azimuth Tables for Facilitating the Determination of Lines of Position and Geographical Position at Sea.*
8. Russell, "On the Navigation of Airplanes," p. 145 for the quotation; Hughes, *History of Air Navigation,* p. 131 for the interpolation. See also Charles Lane Poor, "Recent Advances in the Art of Navigation."
9. Russell, "On the Navigation of Airplanes," p. 149.
10. James Percy Ault, "Navigation of Aircraft by Astronomical Methods," p. 337.
11. Richard Evelyn Byrd, *Skyward,* pp. 62, 76, 81; Eaton, "Aerial Navigation and Navigating Instruments," pp. 786–87.
12. Dobson, "Design of Instruments for Navigation," pp. 381–82; Frederick Tymms, re-

marks on previous paper, p. 388; A. P. Rowe, "Aerial Navigation," pp. 455–56; Harry Egerton Wimperis, "Some Recent Developments in Aircraft Instruments," pp. 5–6; [Plath, C.], *C. Plath,* pp. 112–15; Brackley, *Brackles,* p. 81. Duval and Hébrard, *Traité pratique,* pp. 48–49, after discounting the French gyro sextants, the Baker double horizon sextant, and the older bubble sextants designed for balloons, recommended that seaplanes "alight on the water to take observations." They wrote in 1920 and were not familiar with the postwar American and British bubble sextants.

13. Herbert Nelson Eaton, "Aerial Navigation," *U.S. Air Service* 8 (Oct. 1923): 42; Eaton, "Aerial Navigation and Navigating Instruments," pp. 786–88; P. V. H. Weems, Bubble Sextants, undated MS, WRes, pp. 4–5; Great Britain, Aeronautical Research Committee, "Report for the Year 1920–21," p. 285 for the quotation.

14. E. M. Maitland, "Log of the Pioneer Dirigible Trip," pp. 55, 79; Maitland, "The Transatlantic Voyage of R 34," p. 909; G. H. Scott, "Airship Piloting," p. 1257.

15. Thomas Yeomans Baker and Louis Napoleon George Filon, *Position Fixing in Aircraft during Long Distance Flights over the Sea,* pp. 4–5; Thomas Yeomans Baker, "The Baker Air Sextant"; Dobson, "Design of Instruments for Navigation," p. 381; U.S., War Dept., *Air Service Training Regulation 1440-50, Aircraft Instruments,* pp. 130–31; Harry George Hawker and Kenneth Mackenzie Grieve, *Our Atlantic Attempt,* p. 102.

16. Arthur Whitten Brown, assisted by Capt. Alan Bott, *Flying the Atlantic in Sixteen Hours,* pp. 18–19, 27–30, 44–45, 69, 93; Graham Wallace, *The Flight of Alcock & Brown,* p. 153. The latter, based on materials gathered researching a British Lions film, is well written; but no sources are given.

17. Brown, *Flying the Atlantic,* pp. 42, 46, 51–52, 55–56, 65–76; "The *Daily Mail* £10,000 Atlantic Prize," remarks by Brown at Savoy luncheon, 20 June, *Flight* 11 (26 June 1919): 849.

18. Brown, *Flying the Atlantic,* p. 44; Baker and Filon, *Position Fixing,* pp. 3, 5–9; Wimperis, *Primer of Air Navigation,* pp. 104–07; Wallace, *Flight of Alcock & Brown,* pp. 234, 244. On the difficulties of celestial navigation in an open cockpit, see also Charles Dixon, *The Conquest of the Atlantic by Air,* p. 55, and Russell, "On the Navigation of Airplanes," p. 130.

19. A. C. Read, "The Lame Duck Wins," in G. C. Westervelt, H. C. Richardson, and A. C. Read, *The Triumph of the NC's,* pp. 173, 1, 6–79, 204, 218; "The Cockpit of the Transatlantic Seaplanes," *Scientific American* 120 (24 May 1919): 547, 560; J. C. Hunsaker, "How the NC Boats Were Equipped

for the Pioneer Flight," pp. 172–74.
20. Editorial, *Flight* 11 (29 May 1919): 690; John H. Towers, "The Great Hop," p. 74.
21. Eaton, "Aerial Navigation and Navigating Instruments," p. 800; Towers, "Great Hop," p. 74; Read, "Lame Duck Wins," in Westervelt, Richardson, and Read, *Triumph of the NC's,* pp. 217–18; *Scientific American* 120 (31 May 1919): 588; Walter Hinton, "First Trans-Atlantic Flight," p. 95.
22. Wimperis, "Some Recent Developments," pp. 4–7; Graham Wallace, *R.A.F. Biggin Hill,* pp. 67–68; "H. Hughes & Sons, Ltd.," *Air Annual of the British Empire,* 1929, p. 647; Hughes, *History of Air Navigation,* p. 54; Great Britain, Air Ministry, *Air Navigation Instruments,* p. 27. For the theory of the slide rule, see *Royal Aeronautical Society, Journal* 28 (Feb. 1924): 45–48. For a summary of the various methods available for reducing and plotting celestial observations, see Eaton, "Aerial Navigation and Navigating Instruments," pp. 791–99.
23. Gago Coutinho and Jorge de Castilho, "Navigation des avions du service commercial pour l'Amérique du Sud," pp. 101, 104, 117, 121–24; "H. Hughes & Sons, Ltd.," *Air Annual of the British Empire,* 1929, p. 647; *Aeronautical Digest* 1 (Aug. 1922): 8; Francis M. Rogers, *Precision Astrolabe,* pp. 192–96, 235.
24. Volmerange, "La navigation aérienne en France," p. 749;

William Walter Warlick, *Naval Aviation; A Textbook,* pp. 39–40.
25. Albert B. Duval, "L'emploi des méthodes de navigation est indispensable au succès des transports aériens commerciaux," p. 765.
26. M. O'Gorman, "Notes on Some Scientific and Technical Aspects of the Imperial Air Mail Service," in Great Britain, Civil Aviation Advisory Board, "First Report on Imperial Air Mail Services," p. 28 for the quotation; Great Britain, Advisory Committee on Civil Aviation, "Report on Governmental Assistance for the Development of Civil Aviation," 20 Apr. 1920, p. 4; Great Britain, Air Ministry, "Half-Yearly Report on the Progress of Civil Aviation (Oct. 1, 1919–Mar. 31, 1920)," p. 815; Great Britain, Air Ministry, "Proceedings of the Air Conference, 1920," p. 9; Sefton Brancker, "Aerial Transport from the Business Point of View"; Editorial, *Flight* 12 (18 Nov. 1920): 1186.
27. Sykes, *From Many Angles,* p. 279. The Allied and Associated Powers had signed the *Convention,* the U.S. with reservations, by 31 May 1920. In both titles, "navigation" has the same meaning as in the old British Navigation Acts. The ICAN's title has since been changed to the International Civil Aviation Organization.
28. "International Air Navigation Convention," *Flight* 11 (31 July 1919): 1030. The *Convention,* corrected to include the

many amendments, was published periodically in the International Commission for Air Navigation, *Official Bulletin.*

29. Great Britain, Air Ministry, "Report on the Progress of Civil Aviation (Apr. 1–Dec. 31, 1926)," p. 17; Great Britain, Air Ministry, "Notice to Airmen" (henceforth NOTAM) 29 of 1920, 13 of 1921, 72 of 1922, 36 of 1925, and 82 of 1926, reprinted in *Flight* 12: 374–77, 13: 77, 14: 425, 17: 469, and 18: 864; "British Aerial Navigators' Certificates," *Aerial Age Weekly* 12 (27 Sep. 1920): 83–85; International Commission for Air Navigation, *Official Bulletin* no. 9, pp. 23–24; no. 11, pp. 22–23; no. 13, p. 58; no. 14, p. 25; and no. 15, p. 43; Oliver Stewart, *Flying as a Career,* p. 43 for the quotation.

30. International Commission for Air Navigation, *Official Bulletin* no. 13, p. 48 for the quotation; also no. 2, p. 32; NOTAM 140 of 1920, *Flight* 13 (6 Jan. 1921): 8.

31. Albert B. Duval, "Maps and Navigation Methods," p. 198 for the quotation; Frank Searle, "The Requirements and Difficulties of Air Transport."

32. *Flight* 14 (13, 20 Apr. 1922): 215, 233; NOTAM 64 of 1922, 23 of 1923, and 42 of 1924, in *Flight* 14 (6 July 1922): 386; 15 (5 Apr. 1923): 184; and 16 (29 May 1924): 350.

33. Canada, Department of National Defence, *Report on Civil Aviation,* 1924, p. 23; *Aero Digest* 10 (Jan. 1927): 52.

34. *Air Commerce Bulletin* 1 (16 Dec. 1929): 3; 2 (16 Mar. 1931): 459–64.

35. International Commission for Air Navigation, *Official Bulletin* no. 4, pp. 45–50; no. 20, pp. 43–46; no. 21, pp. 64–65; no. 23, p. 50; no. 26, p. 53; "International Aeronautical Maps," *Aerial Age Weekly* 12 (28 Feb. 1921): 638.

36. "Woodrow Wilson Airway Map Complete," *Aerial Age Weekly* 7 (19 Aug. 1918): 1120–21; *Air Service News Letter* 4 (29 Dec. 1920): 2; 5 (3 Feb. 1921): 1; "Photographic Mapping of Air Routes; America to Make a Practical Start," *Flight* 13 (10 Feb. 1921): 99; Ristow, *Aviation Cartography,* pp. 17–20.

37. U.S., Federal Board of Surveys and Maps, "Report of the Committee on Aerial Navigation Maps," 8 Jan. 1929, adopted 12 Nov. 1929; *Air Commerce Bulletin* 1 (15 Aug. 1929): 21, (2 Dec. 1929): 17, 27–28; 2 (1 Aug. 1930): 70, (15 Jan. 1931): 357–62; 4 (1 June 1933): 587–90; 8 (Dec. 1936): 137; H. F. Ranney, "Maps Again Have Caught Up with Progress"; Raymond L. Ross, "Maps for Aviators."

38. B. Melville Jones, "Flying over Clouds in Relation to Commercial Aeronautics," p. 298 for the quotation; Claude Grahame-White and Harry Harper, *Our First Airways,* illus. opp. p. 76; Lord Montagu of Beaulieu, "The World's Air Routes and Their Regulation"; Frederick Hugh Sykes, "Commercial Avi-

ation in the Light of War Experience," p. 88; Editorial, *Flight* 11 (6 Mar. 1919): 294.
39. *Air Service News Letter* 5 (21 Feb. 1921): 10; "How to Establish an Airway," *Aerial Age Weekly* 13 (30 May 1921): 276–77; U.S., Navy Dept., Bureau of Aeronautics, *Weekly News Letter* 4 (16 Feb. 1925): 2; D. C. Young, "Aerial Highways and Their Markings."
40. NOTAMs reprinted in *Flight* 11 (13 Nov. 1919): 1488; 13 (17 Feb. 1921): 121; *Air Service News Letter* 2 (3 May 1919): 5; Great Britain, Air Ministry, "Half-Yearly Report on the Progress of Civil Aviation (October 1st, 1920–March 31st, 1921)," p. 6; U.S., Commerce Dept., *Report of the Airway Marking Committee, January 23, 1929*, p. 6.
41. Editorial, *Flight* 10 (12 Dec. 1918): 1398; Story B. Ladd, "Aeronautical Roads"; International Commission for Air Navigation, *Official Bulletin* no. 2, pp. 31–32; Lester J. Maitland, "Markings for American Airways," pp. 327–29; *Air Commerce Bulletin* 1 (15 July 1929): 20–21; 5 (Feb. 1934): 194–95.
42. Wolfgang Langewiesche, *I'll Take the High Road*, p. 123 for the quotation; Roderic Maxwell Hill, *The Baghdad Air Mail*, pp. 19–25, 66, 138–39; Philip Albert G. D. Sassoon, *The Third Route*, pp. 249–52; *Flight* 14 (26 Jan. 1922): 50.
43. Great Britain, Air Ministry, "Half-Yearly Report on the Progress of Civil Aviation (Oct. 1, 1919–Mar. 31, 1920)," p. 815; "Aerial Lighthouses," *Scientific American Monthly* 2 (Oct. 1920): 170.
44. R. Preston Wentworth, "The Lighthouse for Aerial Navigation," p. 17; *Air Service News Letter* 5 (9 Nov. 1921): 4; F. A. Collins, "Land Lighthouses"; Great Britain, Air Ministry, "Synopsis of Progress of Work in the Department of Civil Aviation, May 1, 1919–Oct. 31, 1919," p. 6; *Flight* 11: 1442; 12: 23, 124–25, 221, 295, 340, 1186; 13: 210, 252, 280–83, 832, 836; 14: 221, 322; 15: 461.
45. Lyman J. Briggs, "Aeronautical Research at the National Bureau of Standards," p. 485; P. van Braam van Vloten, "L'éclairage des routes aériennes"; Karl F. Löwe, *Flugzeugortung*, p. 55; "Aerial Lighthouses," *Aerial Age Weekly* 14 (20 Feb. 1922): 564–65; *Flight* 17 (30 Apr. 1925): 264.
46. *Air Annual of the British Empire*, 1929, pp. 155–56; F. C. Hingsburg, "Air Navigation Facilities," p. 69; U.S., Commerce Dept., *General Airway Information*, 1 Sep. 1932, pp. 6–8, 68; W. A. Pennow, "Marking the Sky Trails."
47. Hill, *Baghdad Air Mail*, p. 73; *Aeronautical Digest* 3 (Oct. 1923): 227.
48. Frank T. Courtney, "The Future of Air Navigation," p. 265 for the quotation; Montagu of Beaulieu, "World's Air Routes," p. 657; Great Britain, Advisory Committee for Aeronautics, "Report for the Year 1919–20,"

p. 40; M. Deniélou, "T. S. F. et navigation aérienne," p. 234; "Night Flying," *Aeronautical Digest* 1 (Aug. 1922): 10.

49. B. M. Jones, "Flying over Clouds," p. 299; Great Britain, Air Ministry, "Half-Yearly Report on the Progress of Civil Aviation (Oct. 1, 1919–Mar. 31, 1920)," pp. 814, 816; F. W. Dunmore, "Radio Beacons Non-Directive and Directive," p. 1554; J. H. Dellinger and Haraden Pratt, "Development of Radio Aids to Air Navigation," p. 894; R. L. Smith-Rose, "The Equi-Signal Zone Radio Beacon and Air Navigation," p. 98.

50. Great Britain, Air Ministry, "Half-Yearly Report on the Progress of Civil Aviation (April 1st–September 30th, 1921)," p. 7; two later reports in the same series, Oct. 1, 1921–Mar. 31, 1922, pp. 5–6, and Apr. 1, 1925–Mar. 31, 1926, p. 20; *Flight* 13: 158; 14: 299, 304; 17: 718; W. Thomson Lees, "How Radio Serves Aviation Abroad," p. 44; C. B. Carr, "Air Route Radio Services in Great Britain"; International Commission for Air Navigation, *Official Bulletin* no. 19, p. 131.

51. Bellini, "Errors of Direction Finders"; Adcock, "Some Early Observations"; R. L. Smith-Rose, "Radio Direction Finding by Transmission and Reception," pp. 532, 568; *Air Annual of the British Empire*, 1931–32, p. 216; "Wireless and Night Flying; the Marconi–Adcock Direction Finder," *Flight* 24 (19 Feb. 1932): 155; *Air Commerce Bulletin* 2 (15 Nov. 1930): 259–60; 4 (15 July 1932): 33–45, (15 May 1933): 560; Albert F. Hegenberger, "Avigation," p. 219.

52. NOTAM 69 of 1921, *Flight* 13 (6 Oct. 1921): 664.

53. *Nature* 121 (4 Feb. 1928): 189.

54. Great Britain, Air Ministry, "Half-Yearly Report on the Progress of Civil Aviation (October 1st, 1920–March 31st, 1921)," p. 6; H. C. Fleming, "How the ZR 3 Was Piloted by Radio," p. 1135; C. E. Horton and C. Crampton, "Radio Compass Developed in H. M. Signal School," pp. 284–86; *Air Annual of the British Empire*, 1935–36, p. 451.

55. Wallace, *R. A. F. Biggin Hill*, p. 31, claims the equipment on the NCs "had been placed at the disposal of the American government by the Royal Air Force," but he gives no sources. Edgar H. Felix, "Aircraft Radio Direction Finding Equipment Developed by the Navy Department," p. 852, says the USN Bureau of Steam Engineering "developed" the instruments.

56. Charles I. Stanton, paper read at the 9th annual meeting of the Institute of Navigation, New London, Conn., 28 Aug. 1953, copy at WRes; "One Year of Air Mail Service," *Scientific American Supplement* 88 (2 Aug. 1919): 79; *Air Service News Letter* 4 (14 Sep. 1920): 1–2, (25 Sep. 1920): 1; 5 (12 Jan. 1921): 2; U.S. Navy, *Daily Aviation News Bulletin*, 7–8

July 1920; James C. Edgerton, "Radio as Applied to Air Navigation in the Air Mail Service."
57. O. Scheller, "Flugzeugsteuerung bei unsichtigem Wetter"; Walter, *Directive Wireless Telegraphy,* pp. 113–15; Smith-Rose, "Equi-Signal Zone," p. 98; A. M. Jacobs, "Radio Beacon Guides Night Air Mail"; Dunmore, "Radio Beacons Non-Directive and Directive"; William H. Murphy, "The Interlocking Equisignal Radio Beacon"; Dellinger and Pratt, "Development of Radio Aids," pp. 892–94; "Aircraft Radio Development," *Air Commerce Bulletin* 1 (15 Aug. 1929): 5–6.
58. Dellinger and Pratt, "Development of Radio Aids," pp. 892, 894.
59. Ibid. The British allowed private aircraft to use their direction finding net after Dec. 1926. NOTAM 76 of 1926, *Flight* 18 (2 Dec. 1926): 796.
60. *Air Commerce Bulletin* 4 (15 July 1932): 33–45, (15 May 1933): 560; 9 (July 1937): 1–3; Daniel Guggenheim Fund for the Promotion of Aeronautics, *Equipment Used in Experiments to Solve the Problem of Fog Flying,* pp. 9–10; U.S., Commerce Dept., *General Airway Information,* 1 Sep. 1932, p. 84.
61. *Air Commerce Bulletin* 5 (Mar. 1934): 224; 8 (Sep. 1936): 65–70; H. I. Metz, "Cone of Silence Markers Identify Exact Locations of Range Stations," pp. 169–70; E. A. Cutrell, "Instrument and Radio Flying," p. 14;

P. T. W. Scott, "Flight Log," pp. 18, 20.
62. George Kimball Burgess, "Aircraft Radio Beacon Development," p. 1764; *Air Commerce Bulletin* 2 (2 Mar. 1931): 437–38; 5 (Sep. 1933): 89–90.
63. G. W. Williamson, "Instrument Progress and Development," p. 177; George R. Fairlamb, Jr., "Air Navigation; an Estimate of Progress Made," p. 151.
64. Cy Caldwell, "Getting Lost in Fog or Stormy Weather," p. 27 for the quotation; *Aviation* 36 (Feb. 1937): 52–53; 37 (May 1938): 20.
65. Henri Busignies, "Appareils indicateurs donnant par lecture directe la direction d'une onde"; P. Franck, "Le radiocompas et la navigation aérienne"; Great Britain, Air Ministry, "Report on the Progress of Civil Aviation (April 1–December 31, 1926)," p. 25; *Air Commerce Bulletin* 1 (2 June 1930): 4; 3 (15 Oct. 1931): 187–88; 4 (15 May 1933): 559; 5 (Aug. 1933): 33; 7 (Jan. 1936): 163–64.
66. Frederick W. Lutz, "Kreusi Radio Compass for Commercial Use"; [Arthur Hughes ?], Conference with Fairchild Officials, 15 May 1935, unsigned, typewritten notes, WRes; John P. Gaty, "Radio Compass"; Richard Duncan, *Air Navigation and Meteorology,* pp. 210–14; *Air Annual of the British Empire,* 1935–36, pp. 449–50, 458; *Aviation* 35 (Feb. 1936): 50; Don Fink, "Coast Guard Radio," pp. 48–49; "An Auto-

matic Direction Finder," *Communications* 18 (Oct. 1938): 10–11.
67. Roderick Denman, "Radio Air Navigation," p. 54 for the quotation; David Hay Surgeoner, *Aircraft Radio,* p. 3; C. H. McIntosh, "Navigating with the D/F Loop," pp. 52–53.
68. David G. C. Luck, "Omnidirectional Radio Range."
69. Frederick Tymms, "Practical Navigation of Aircraft," pp. 202–03 for the quotations; Duval and Hébrard, *Traité pratique,* pp. 22, 28; Harold Gatty, commenting on paper by George R. Fairlamb, Jr., *Aeronautical Engineering* 1 (July–Sep. 1929): 155; Harold Gatty, "Aerial Navigation; Methods and Equipment," p. 154; F. Bardot, "Guidage radioélectrique des aéronefs," p. 282.
70. P. V. H. Weems, "Air Navigation; Finding Your Way in the Air," *Aviation* 36 (June 1937): 76; Norman Macmillan, ed., *Let Experts Tell You; How We Fly,* pp. 95–96.

CHAPTER 6: TRANSOCEANIC AIRLINES AND CELESTIAL NAVIGATION

1. R. E. G. Davies, *A History of the World's Airlines,* pp. 144–48, 218–22, 320–24.
2. Nevil Shute [Nevil Shute Norway], *Slide Rule,* pp. 200–02; *Aeroplane* 47 (26 Dec. 1934): 779; Paul L. Briand, Jr., *Daughter of the Sky,* pp. 191–99; Amelia Earhart, *Last Flight,* pp. 135–36.
3. Charles Edward Kingsford-Smith and Charles T. P. Ulm, *The Flight of the Southern Cross,* pp. 78–79; Fairlamb, "Air Navigation," p. 153.
4. *Aviation* 24 (5 Mar. 1928): 581; 34 (July 1935): 55; *Air Commerce Bulletin* 3 (15 Oct. 1931): 188; 4 (15 May 1933): 559; 6 (Apr. 1935): 215; William Stephen Grooch, *Skyway to Asia,* pp. 75–76, 81, 176–78, 193–94; Matthew Josephson, *Empire of the Air,* pp. 40–41. For Imperial Airway's use of radio, John H. Millar, "Radio on the Empire Routes." For the problems of plotting long-range radio bearings, L. Driencourt, "Le planisphère de Mercator."
5. "The Army Flight to Hawaii," *Aero Digest* 11 (July 1927): 16; Albert F. Hegenberger and Bradley Jones, "Oversea Navigation," p. 702; Clayton C. Shangraw, "Radiobeacons for Trans-Pacific Flights."
6. Franklin L. Hunt, "Recent Developments and Outstanding Problems," p. 814.
7. Mary Heath, "The Technical Aspects of a Trans-African Flight on a Light Aeroplane," p. 27; George Tate, Jr., "Improving Airspeed Accuracy by Pitot Tube Location"; F. L. Thompson, "The Measurement of Air Speed of Airplanes"; Georg Kiel, "Measurement of the True Dynamic and Static Pressures in Flight"; Williamson, "Instrument Progress and Development," pp. 187–88; Thomas L. Thurlow to Weems, 11 Jan. 1940, WML, Thurlow.
8. W. Möller, "Kompassübertra-

gungen und Kompasssteuerungen im Flugzeug," pp. 62–63.
9. Warner, "Direction Instruments," pp. 640–41; Paul R. Heyl and Lyman J. Briggs, "The Earth Inductor Compass"; George W. Polk, "Instruments Used by the World Flyers," p. 34; U.S., Navy Dept., Bureau of Aeronautics, *Weekly News Letter* 3 (16 June 1924): 4–5; Lowell Jackson Thomas, *The First World Flight*, p. 7; *Automotive Industries* 52 (7 May 1925): 829; Victor E. Carbonara, "Aircraft Instruments for Oversea Navigation," p. 528; Brice Goldsborough, "The Earth Inductor Compass."
10. Charles A. Lindbergh, "My Flight to Paris," pp. 516–18 for the quotation; Clarence Duncan Chamberlain, *Record Flights*, p. 61.
11. Byrd, *Skyward*, pp. 266–70; *L'année aéronautique*, 1927–28, pp. 231–32. Bernt Balchen, *Come North with Me*, pp. 114–15, does not mention compass difficulties and blames their failure to reach Paris on Byrd's refusal to follow Balchen's advice. In view of the publication dates of the two books, the general egotism pervading Balchen's, and a conversation with an expert on aerial navigation who knew both men, I prefer Byrd's account.
12. *L'année aéronautique*, 1928–29, pp. 267–68; *Aviation* 24 (30 Apr. 1928): 1230; Kingsford-Smith and Ulm, *Flight of the Southern Cross*, pp. 241–42.
13. Douglas Corrigan, *That's My Story*, p. 94; Charles A. Lindbergh, *The Spirit of St. Louis*, pp. 336–37. How to account for the discrepancy? The earlier account normally commands the historian's respect, but in this case I prefer the later. The young hero was immediately pressed to describe his achievement in print. He had succeeded brilliantly, with a near-perfect landfall in Ireland and a safe landing at Le Bourget with near-dry tanks. Both the *Aero Digest* article (note 10) and *"We"; The Famous Flier's Own Story*, were hurriedly written, and he was in no mood to dwell on the difficulties. In later life Lindbergh developed an extreme dislike for inaccuracies in print, and he wrote *Spirit of St. Louis* largely to set the record straight.
14. J. D. Tear, "Electromagnetic Compass Systems for Aircraft," pp. 471–72; Möller, "Kompassübertragungen und Kompasssteuerungen," p. 64; Mortimer F. Bates, "Filling Up the Instrument Panel," p. 196.
15. U.S., Air Service, Engineering Division, *Technical Orders* no. 14, Mar. 1920, p. 12; *Air Service News Letter* 6 (3 May 1922): 2; "Equipment Development at McCook Field," *Aerial Age* 16 (June 1923): 289; Hubert Wilkins, "Some Aspects of the Graf Zeppelin's Flight around the World," pp. 597, 600; *Aero Digest* 20 (Jan. 1932): 58; Briand, *Daughter of the Sky*, p. 83; P. V. H. Weems, "The Flight of 'Tingmissartoq,'" pp. 35, 102; Anne

Morrow Lindbergh, *Listen! the Wind*, p. 264.

16. "Sperry's Spinning Wheel," *Fortune* 21 (May 1940): 121; Williamson, "Instrument Progress," p. 195; Brig. Gen. A. W. Robins, "Materiel Division Activities for the Year 1937"; U.S., Navy Dept., *Aircraft Navigation Manual*, H. O. pub. 216 (1940), pp. 73–74.

17. Burt M. McConnell, "The Aeroplane in Arctic Exploration"; "The Woodward Aero Navigator," *Aerial Age Weekly* 5 (26 Mar. 1917): 51.

18. Byrd's dispatch of 5 Aug. 1925 is in U.S., Navy, Bureau of Aeronautics, *Weekly News Letter* 4 (10 Aug. 1925): 4. For the Bumstead compass, R. E. Byrd, "Navigation Problems of the MacMillan Expedition"; Nell Ray Clarke, "Are We over the Pole?"; "Commander Byrd's New-Type Compass," *Flight* 21 (27 Dec. 1929): 1349; R. E. Byrd, *Little America*, pp. 310, 314. For the Goerz compass, C. J. Stewart, "Determination of Position in High Latitudes," pp. 212–14; *Scientific American* 131 (Nov. 1924): 332–33.

19. Hj. Riiser-Larsen, "The Navigation over the Polar Sea," in Roald E. G. Amundsen and Lincoln Ellsworth, *First Crossing of the Polar Sea*, pp. 192–93 for the quotation; Byrd, *Skyward*, p. 192; George Baidukov, *Over the North Pole*, pp. 33, 42, 57, 60.

20. Hans Boykow, "Probleme der terrestrischen Navigation im Luftfahrzeuge," p. 34 for the quotation; Riiser-Larsen, "The Navigation over the Polar Sea," in Amundsen and Ellsworth, *First Crossing of the Polar Sea*, pp. 183–87; Wilhelm Niemann, "Langstrecken-Navigation in Luftfahrzeugen über Land und See im In- und Ausland," pp. 1045–46; Löwe, *Flugzeugortung*, pp. 75–78.

21. Yves le Prieur, "Le correcteur de route"; Eaton, "Aerial Navigation and Navigating Instruments," pp. 777–79; *Scientific American Monthly* 3 (May 1921): 453–54; U.S., War Dept., Air Service Training Regulation 1440–50, *Aircraft Instruments*, 23 Jan. 1925, pp. 104–07; Yves le Prieur, "Avec la mission de Goys"; Pierre Léglise, "Notes techniques sur une série de performances d'hydraviation marchande."

22. U.S., Navy Dept., Bureau of Aeronautics, *Weekly News Letter* 3 (10 May 1924): 1; Franklin L. Hunt, "New Types of Aircraft Instruments," pp. 239–41; Lyman James Briggs and others, "Present Status of Aircraft Instruments," p. 155.

23. "Gatty's Navigation Instrument," *Aero Digest* 20 (Jan. 1932): 57–58; P. V. H. Weems, "Gatty Ground Speed Meter"; Gatty, "Aerial Navigation," p. 155; Weems to Purchasing Agent, Transcontinental and Western Airline, Kansas City, 13 Feb. 1934, WML, Pioneer. Evolution of Gatty's instrument may be traced in Weems System of Navigation Catalogs, WML.

24. H. M. Cave-Browne-Cave, "Cruise of the Royal Air Force

Far East Flight," p. 565; C. L. Scott, "By Flying Boat to India," pp. 823–27; S. Paul Johnston, "Clipper Gear," pp. 15–16. For a comparison of contemporary driftmeters, L. Aussenac, "Les cinémodérivomètres modernes."

25. Robins, "Materiel Division Activities for the Year 1937."
26. *Aviation* 27 (2 Nov. 1929): 917; Daniel Guggenheim Fund, *Equipment Used in Experiments to Solve the Problem of Fog Flying*, pp. 51–53; Robins, "Materiel Division Activities for the Year 1937"; Williamson, "Instrument Progress."
27. H. White-Smith, "The Operation of Civil Aircraft in Relation to the Constructor," p. 35 for the quotation; E. L. Ellington, "The Present Position of Aircraft Research and Contemplated Developments," pp. 64–65; Lawrence A. Hyland, "True Altitude Meters," pp. 1356–58.
28. Shute, *Slide Rule*, pp. 102–04.
29. *Scientific American* 133 (Aug. 1925): 129. Ludwig Dürr, *Fünfundzwanzig Jahre Zeppelin-Luftschiffbau*, describes the instruments installed in the *LZ 126* but does not mention absolute altimeters, possibly because they were experimental. See also Hugo Eckener, *My Zeppelins*, p. 27.
30. Hyland, "True Altitude Meters," pp. 1322–23; Wilkins, "Some Aspects of the Graf Zeppelin's Flight," p. 597; Heinrich Koppe, *Luftnavigierung und die Arbeiten des Navigierungs-Ausshusses der WGL*, p. 4; Hughes, *History of Air Navigation*, pp. 67–68; *Air Corps News Letter* 16 (25 Jan. 1932): 14; *Aero Digest* 24 (Jan. 1934): 49; William G. Brombacher, "Measurement of Altitude in Blind Flying," p. 1; C. S. Draper, "Sonic Altimeter for Aircraft"; Leo P. Delsasso, "The Acoustic Altimeter."
31. Hyland, "True Altitude Meters," pp. 1323, 1356; E. F. W. Alexanderson, "Height of Airplane above Ground by Radio Echo," pp. 616–17; *Aviation* 26 (20 Apr. 1929): 1337; Sadahiro Matsuo, "Altimètre à lecture directe pour l'aéronautique"; Leonard H. Engel, "Principle of the Echo Altimeter"; Don Fink, "Aviation Radio," a column, *Aviation* 37 (Nov. 1938): 48–49, 80; Lloyd Espenschied and R. C. Newhouse, "A Terrain Clearance Indicator," pp. 222–34; *Aviation* 38 (Dec. 1939): 75; U.S., Navy Dept., *Aircraft Navigation Manual*, H. O. pub. 216 (1940), p. 72.
32. Byrd, *Skyward*, p. 258 for the quotation; Byrd, "Our Transatlantic Flight," p. 8; James H. Kimball, "Problem of Transatlantic Weather Forecasting"; Eckener, *My Zeppelins*, pp. 48–50; Niemann, "Langstrecken-Navigation in Luftfahrzeugen," p. 1082; Daniel Sayre, "First 1/2 Million Miles of Pacific Progress," p. 44; "Tomorrow's Airplane," *Fortune* 18 (July 1938): 55; Donald Clifford Tyndall Bennett, *Pathfinder*, pp. 76–77.
33. Patrick Gordon Taylor, *The Sky Beyond*, p. 25.

34. Weems to P. F. Everitt, Henry Hughes & Son, 6 Aug. 1929, WRes, for the quotation; Weems's comments on paper by George R. Fairlamb, Jr., *Aeronautical Engineering* 1 (July–Sep. 1929): 157; Weems to A. A. Priester, Operation Manager, Pan American Airways, 19 May 1929, WRes; Weems to Hughes, 22 Dec. 1936, WRes.

35. Weems to W. A. Reichel, Pioneer Instrument, 20 Feb. 1934, WML, Pioneer, for the quotation; Kingsford-Smith and Ulm, *Flight of the Southern Cross*, pp. 99, 126; Weems to Reichel, 28 Apr. 1934, WML, Pioneer; Air Corps form 54, Unsatisfactory Report, prepared by Capt. Grandison Gardner, Maxwell Fld., 20 Nov. 1934, in AFArc 248.2016J; U.S., Navy Dept., Bureau of Aeronautics, *Newsletter* 12 (1 Oct. 1935): 7; Baidukov, *Over the North Pole*, pp. 41–42; M. Drodofsky, "Astronomische Beobachtungsinstrumente," p. 7; E. A. Link, Link Aviation, to Peter H. Redpath, Navigation Engineer, Transcontinental & Western Airlines, 2 Dec. 1940, WML, Link, I.

36. Coutinho and Castilho, "Navigation des avions," p. 105; Louis Kahn, "Les méthodes de navigation employées par les aviateurs Costes et Bellonte," p. 470; Weems, "Flight of the 'Tingmissartoq,'" p. 102; "Tomorrow's Airplane," *Fortune* 18 (July 1938): 91.

37. Koppe, *Luftnavigierung*, p. 7; Kahn, "Les méthodes de navigation employées par les aviateurs Costes et Bellonte," p. 470. U.S., Navy Dept., Bureau of Aeronautics, *News Letter* 12 (1 Oct. 1935): 8 recommended a procedure similar to Bellonte's.

38. P. V. H. Weems, "Analyse des erreurs dues aux accélérations de la bulle," p. 78, and Weems to the Secretary of the Navy, 22 Nov. 1935, WRes, for the quotations; Weems, "Accuracy of Bubble Sextant Observations"; Thurlow to Weems, 7 Dec. 1935, WRes; Weems to Victor E. Carbonara, Kollsman Instrument, 9 Oct. 1935, WRes; Carbonara to Weems, 18 Oct. 1935, WRes.

39. Weems to Hughes, 22 Dec. 1936; P. F. Everitt, a Hughes engineer, Averaging Device for Air Sextant, typescript, 24 July 1936; Thurlow to Weems, 29 Jan. 1937; Everitt, Additional Instructions for Averaging Sextant, typescript, 10 Feb. 1937; Thurlow to Weems, 3 Aug. 1937; Instructions for Use of Husun Bubble Sextant Mk. XII, typescript, 26 Oct. 1937; Thurlow to Weems, 7 Nov. 1937; Weems to Everitt, 20 Dec. 1937, all WRes.

40. Carbonara to Weems, 18 Oct. 1935, WRes; Weems to Sqdn. Ldr. John F. Griffiths, RCAF, commanding 99th RAF Sqdn., Mildenhall, Suffolk, 26 May 1938, WML, Astrograph.

41. A. Bastien, "Le point astronomique," pp. 68–69; Hughes, *History of Air Navigation*, pp. 120-28; Link to Weems, 10 May 1938, WML, Link, I; Thurlow to Weems, 6 Dec. 1938, WRes.

42. Affidavit dated 7 Feb. 1939, WRes; Weems to Link, 15 Feb. 1939, WML, Link, I; Link to Weems, 18 Feb. 1939, ibid. for the quotation.
43. Thurlow to Weems, 15 Mar. and 29 May 1939, WRes; G. A. Lewthwaite, Asst. Sales Manager, Pioneer, to Weems, 6 Mar. 1940, WML, Pioneer; Weems to Francis Chichester, Hughes & Son, 9 Apr. 1940, WML, Husun, VII; Thurlow to Weems, undated, about 1 Mar., and 12 Oct. 1940, WML, Thurlow; Link to Weems, 4 Mar. and 25 Oct. 1941, WML, Link, II; "New Link Octant and Collimator," *Aviation* 40 (Sep. 1941): 84.
44. Hughes, *History of Air Navigation*, pp. 121–29; Thurlow to Weems, 3 Feb. 1941, WML, Thurlow.
45. Thurlow to Weems, 3 and 11 Jan. 1940, WML, Thurlow. The error introduced into celestial observations by Coriolis's acceleration was described by M. Drodofsky, "Astronomische Beobachtungsinstrumente," Germany, Reichsluftfahrtministerium, *Ringbuch der Luftfahrttechnik*, Section VB2, p. 4, 30 Sep. 1937. This may have been Stewart's source.
46. Weems–Thurlow correspondence, 3 Jan.–13 Apr. 1940, WML, Thurlow; Weems to Griffiths, 14 Feb. 1940, WML, Astrograph; Weems to Gunne Lowkrantz, Link Aviation, 2 Apr. 1940, WML, Link, I; Reichel to Weems, 3 May 1940, WML, Pioneer. See also Weems and Thurlow, "Earth's Rotational Effect on the Bubble Sextant," and John Q. Stewart, "Correction Table for the Effect of Coriolis Acceleration on Bubble Sextant Altitudes."
47. Zeh Bouck, "W2XBQ Flies to Bermuda," p. 72; Sqdn. Ldr. David J. Waghorn, Hq. #4 Bomber Gp., Linton-on-Ouse, York, to Weems, 1 Aug. 1938, WML, Astrograph.
48. C. Plath and F. Schilly, "Sextant à périscope avec compas associé"; [C. Plath, firm], *C. Plath*, pp. 127–28. Gustav Förstner of the DVL wrote two technical reports on the sextant. An abstract of #349 is in *Aero Digest* 25 (July 1934): 29; #356 is in *Zeitschrift für Flugtechnik und Motorluftschiffahrt* 24 (28 Dec. 1933): 680–84.
49. Thomas L. Thurlow, "Navigation of the Howard Hughes Around the World Flight," p. 1 for the quotation; Earhart, *Last Flight*, p. 54; Daniel Sayre, "Pacific Preview," p. 82; Johnston, "Clipper Gear," p. 16; "Tomorrow's Airplane," *Fortune* 18 (July 1938): 52; Boeing Aircraft Company, *Handbook of Instructions for Operation of the Boeing Model 314 Flying Boat*, p. 31.
50. Weems to Link, 4 Apr. 1938, WML, Link, I; Weems to Waghorn, 12 Aug. 1938, WML, Astrograph; *Aviation* 37 (Jan. 1938): 31. For later USAAF astrodomes, Conference with Maj. Thurlow, in Diary of Conference with Representatives of Materiel Division, 8–9 June 1941, by Maj. Harbold . . . of the Air Navigation Training

School, Barksdale Fld., La., and Capt. O. E. Henderson (Office, Chief of the Air Corps) to Maj. N. B. Harbold, 30 June 1941, both in AFArc 220.7193-8, v. 1.

51. Weems to Carl L. Bausch, Bausch & Lomb Optical, 4 Nov. 1929, WRes, for the quotation; Weems to E. B. Collins, Hydrographic Office, 20 July 1929, WML, Hydrographic Office; Hughes, *History of Air Navigation*, pp. 71–72.

52. Weems to Collins, 18 Jan. 1933, WRes; *Aero Digest* 22 (Jan. 1933): 36; Weems, "Flight of the 'Tingmissartoq,'" p. 105 for the quotation; Weems to Chichester, 5 Jan. 1940, WML, Husun, VII; A. Bastide, Review of *Éphémérides aéronautiques pour 1937*, the second French almanac, the first having been published for 1936.

53. *Flight* 32 (30 Sep. 1937): 336; Hughes, *History of Air Navigation*, pp. 72–73.

54. Weems to Lt. J. C. Foust, H. O., 13 Sep. 1939, WML, Hydrographic Office; Weems to Thurlow, 24 May 1940, WML, Thurlow.

55. E. B. Collins to Weems, 15 June 1939; Weems to Collins, 16 June 1939; Collins to Weems, 26 June 1939; J. Y. Dreisonstok (H. O. representative at the Naval Observatory) to Weems, 25 Nov. 1939, all in WML, Hydrographic Office.

56. Thurlow to Weems, undated, about 28 May 1940, WML, Thurlow, for the quotation; G. M. Clemence and D. H. Sadler, "Unification of the Air Almanac and the American Air Almanac."

57. Francis Charles Chichester, *Seaplane Solo*, p. 64; Chichester, *Ride on the Wind*, p. 247; Bennett, *Pathfinder*, p. 69.

58. Hughes, *History of Air Navigation*, p. 71, and Collins to Weems, 2 Jan. 1929, WML, Hydrographic Office, for the quotations; Weems, *Line of Position Book;* Gatty, "Aerial Navigation," pp. 156–58; A Discussion of the Article, Line of Position, by Mr. E. B. Collins, Nautical Expert, H. O., May 1927, mimeographed, and Weems to the Hydrographer, 13 Mar. 1929, both WML, Hydrographic Office; Weems to Charles H. Colvin, General Manager, Pioneer Instrument, 4 Aug. 1928, and Weems to Albert McGall, Orange, N.J., 22 Sep. 1929, both WML, Pioneer; Weems to Everitt, 5 May 1931, WRes; U.S., Navy Dept., Hydrographic Office, *Navigation Tables for Mariners and Aviators: Dreisonstok*, H. O. pub. 208 (1928); John Edward Gingrich, *Aerial and Marine Navigation Tables;* U.S., Navy Dept., Hydrographic Office, *Tables of Computed Altitude and Azimuth*, H. O. pub. 214 (9 vols., 1936–46); William Caveny Eberle and P. V. H. Weems, *Learning to Navigate*, p. 49.

59. For a critical comparison of nine contemporary reduction methods, A. Bastide, "De quelques procédés récents de points astronomiques en vol." For simple errors after hours in the

air, Thurlow, "Navigation of the Howard Hughes Around the World Flight," p. 2, and "Wing Commander," *Bombers' Battle,* p. 35.
60. P. V. H. Weems, *Star-Altitude Curves for Obtaining a Fix . . . Latitude 30° to 41° North;* Weems to Chichester, 8 Dec. 1939, WML, Husun, VII; K. Hilding Beij, "Astronomical Methods in Aerial Navigation."
61. Weems's patent, 2,143,042, in WML, Patents, was awarded 10 Jan. 1939; application date was 31 July 1929; Chichester, *Seaplane Solo,* pp. 67–71; Flight Cadet Under Officer P. G. St. G. O'B., "Colonel Lindbergh's Lecture," *Royal Air Force College Journal* 18 (Spring 1938): 47.
62. Wiley Post and Harold Gatty, *Around the World in Eight Days,* pp. 97, 115, 211; Everitt to Weems, 31 Oct. 1929, WRes.
63. Kahn, "Méthodes de navigation employées par les aviateurs Costes et Bellonte"; Ardouin-Dumazet, "Démonstration élémentaire de propriétés des cartes orthodromiques."
64. A. G. von Baumhauer, "Sphero-Triangulator"; Bastide, "De quelques procédés récents de points astronomiques," pp. 95–96; *Aviation* 35 (Nov. 1936): 43; Weems System of Navigation to Pacific Scientific Co., San Francisco, 16 May 1936, WML, Pioneer; Robins, "Materiel Division Activities"; U.S., War Dept., *Celestial Air Navigation,* TM 1-206, 4 Mar. 1941, pp. 88–94.
65. Weems to Thurlow, 3 June 1937, WRes; Thurlow to Weems, undated, about 11 Dec. 1940, WML, Thurlow, for the quotation; John D. Peace, Jr., "Fix Calculator."
66. Griffiths to Weems, 6 July 1938, WML, Astrograph, for the quotation; Guyot, "Un nouvel appareil de calcul rapide du point astronomique à bord d'avion"; "Le calculateur–marqueur de point astronomique Bastide-Lepetit type 300," *Revue de l'Armée de l'Air* 8 (Dec. 1936): 1425–29.
67. Griffiths–Weems correspondence, 9 Oct. 1938–10 Feb. 1939, WML, Astrograph; Thurlow to Capt. O. E. Henderson, T & O Division, Air Corps, Washington, 9 June 1941, WML, Thurlow.
68. Griffiths, RAF Station Mildenhall, Bury St. Edmunds, to Weems, 2 May 1938, WML, Astrograph, for the quotation; "The Americas and the Orient Linked by Air," *Aero Digest* 27 (Dec. 1935): 26; *Aviation* 38 (Mar. 1939): 33; Sayre, "First 1/2 Million Miles of Pacific Progress"; Weems to Link, 20 Apr. 1938, WML, Link, I; Weems to Griffiths, 26 May 1938, WML, Astrograph; Weems to Thurlow, 14 Mar. and 3 June 1940, WML, Thurlow.
69. Thurlow to Weems, 6 June 1941, and Thurlow to Henderson, 9 June 1941, both WML, Thurlow.
70. Two have been omitted. For charts, other than U.S. which are dealt with in Chapter 5, see Ristow, *Aviation Cartog-*

raphy, pp. 21–24. For DR computers: Boykow, "Probleme der terrestrischen Navigation," p. 36; *Flight* 21 (16 May 1929): 403–04; 24 (5 Aug. 1932): 739–40; 25 (7 Sep. 1933): 893; 31 (11 Feb. 1937): 150; *Air Annual of the British Empire, 1934–35,* pp. 640–42; 1938, p. 470; P. V. H. Weems, "Mark VII Navigational Computer for Aircraft"; Weems, "Dalton Air Navigation Computer Type E-1A"; Blondel la Rougery, "Documentation sur les instruments de navigation," pp. 134–35, 159.

71. Sayre, "First 1/2 Million Miles of Pacific Progress," p. 44, and "Tomorrow's Airplane Today," *Fortune* 18 (July 1938): 55, for Pan American; for the quotations, Francis Chichester, "Navigation—Fourth Rate or First Class?," p. 56; A. Bastien, "Long Distance and Night Navigation," pp. 13, 18; Alfred Henke, "Langstreckennavigation," p. 12.

CHAPTER 7: NAVIGATION OF LONG-RANGE BOMBERS IN THE U.S. ARMY AIR CORPS TO 1941

1. Commander Blue Patrol Wing to Blue Commandant, 12th Naval District, subject: Coastal frontier defense joint exercise #4, quoted in History of Langley Field, 1 Mar. 1935–7 Dec. 1941, 1: 105–06, AFArc 285.49-2.
2. See Webster and Frankland, *Strategic Air Offensive against Germany,* 1: 163–253, for an excellent presentation of the British experience. In vol. 4, Annex I, "The Principal Radar Aids to Navigation and Bomb Aiming," Annex III, "Operational Training in Bomber Command," and Appendix 13, "Report by Mr. Butt to Bomber Command on His Examination of Night Photographs, 18th August 1941," are pertinent.
3. Giulio Douhet, *The Command of the Air,* p. 21 for the quotation, also pp. 9, 40–58.
4. "The War of 19—," first published March 1930, ibid., p. 347 for the quotation, also p. 62.
5. Photostat of typescript in Air Force Academy Special Collections. Quotations are from pp. 67, 98, 75.
6. U.S., War Dept., Air Service Tactical School, *Bombardment Course, 1924–1925,* pp. 20, 42 for the quotations, also pp. 7, 27, 56, AFArc 168.69-4; Robert T. Finney, *History of the Air Corps Tactical School, 1920–1940,* pp. 29–31.
7. U.S., War Dept., Air Corps Tactical School, *Bombardment Aviation Course Textbook* (Feb. 1931), pp. 119, 120, 37 for the quotations, AFArc 248.101-9; *Air Corps News Letter* 15 (10 Apr. 1931): 145.
8. Finney, *History of the Air Corps Tactical School,* pp. 32–33.
9. A word used in the early 1930s for "aerial navigation." It was apparently coined by Lts. Lester J. Maitland and Albert F. Hegenberger after their San Francisco–Hawaii flight. U.S.,

Navy Dept., Bureau of Aeronautics, *News Letter* 5 (17 Aug. 1927): 13; and interview with Maj. Gen. Norris B. Harbold (USAF, Ret.), 2 July 1969.
10. U.S., War Dept., Air Corps Tactical School, *Bombardment Textbook* (Nov. 1935), pp. 8, 137, 145, AFArc 248.101–9.
11. Ibid., p. 129; A. P. Rowe, *One Story of Radar*, p. 2.
12. Air Corps Tactical School, 1939–40, Air Force course, lesson 28, Bombardment in counter-air operations, lecture, AFArc 248.2021A–28. Additional correspondence on flares is in 145.91–535.
13. Henry Harley Arnold and Ira Clarence Eaker, *Winged Warfare*, pp. 137–38.
14. "Permanent Organization of the Royal Air Force," Memorandum by the Chief of the Air Staff, 25 Nov. 1919, Cmd. 467, *Parliamentary Papers*, 1919, 33: 5 for the quotation; *Flight* 12 (19 Feb. 1920): 203; 13 (29 Sep. 1921): 652; Flying Officer Gillman, commenting on paper by Frederick Tymms, *Royal Aeronautical Society Journal* 29 (May 1925): 237; Bennett, *Pathfinder*, pp. 47–51.
15. Chichester, "Navigation— Fourth Rate or First Class?," p. 55.
16. U.S., Air Service, Engineering Division, *Technical Orders* no. 8, Sep. 1919, pp. 68–73; *Air Service News Letter* 2 (6 June 1919): 2–3; 4 (16 July 1920): 13; and for the quotation, 5 (15 Mar. 1921): 9.
17. U.S., Navy Dept., *Daily Aviation News Bulletin*, 8 Mar. 1920; *Air Service News Letter* 4 (28 July 1920): 4; and for the quotation, (13 Aug. 1920): 16; "Synopsis of Instruction for the Air Service Primary Flying School, Brooks Field, San Antonio," printed as appendix to Henry Harley Arnold, *Airmen and Aircraft;* History of the AAF Flying Training Command and Its Predecessors, 1 Jan. 1939 to 7 July 1943, 1: 37–50, AFArc 221.01.
18. History of Randolph Field, Texas, 1931–1944, 1: 105 for the quotation, 176 for the figures, AFArc 287.86–1.
19. Proceedings of a Board of Officers, McCook Field, 27 Jan. 1927, AFArc 248.122–5; Memo for Gen. Foulois by Capt. Harold M. McClelland, 2 Oct. 1928, not seen, cited in Ben Baldwin, Individual Training of Navigators in the AAF, U.S. Army Air Forces historical study #27, Jan. 1945, typescript in Air Force Academy Library Special Collections; interview with Maj. Gen. Norris B. Harbold (USAF, Ret.), 2 July 1969.
20. Finney, *History of the Air Corps Tactical School*, pp. 20–22.
21. Form II, ACTS, Air Force course, 4–1, 29 Oct. 1929, a lecture outlines, AFArc 248. 2011A–4; Instructors' file for the Navigation Course, 1934, and for the quotations, Training Directive AN–4, ACTS, 1934–1935, and Memorandum of 20 Dec. 1934, all in AFArc 248.2016J; U.S., War Dept., Air Corps Tactical School, *Aerial*

Navigation (Nov. 1931), AFArc 248.101–13.
22. Instructor's pencil notes for lesson #1, Navigation Course, 1935, AFArc 248.2017J–1. I have bowdlerized the chicken story.
23. Memo for Gen. Pratt from Col. H. A. Dargue, Asst. Commandant, 5 Apr. 1937, subject: The Air Navigation Course, AFArc 248.2018J.
24. *Air Corps News Letter* 15 (10 Apr. 1931): 130; 16 (15 Mar. 1932): 94, (3 May 1932): 156, (27 Aug. 1932): 345; interview with Maj. Gen. Norris B. Harbold (USAF, Ret.), 2 July 1969.
25. Baldwin, Individual Training of Navigators in the AAF, pp. 44–47, 55–56; Norris B. Harbold, *The Log of Air Navigation*, pp. 38 ff.
26. Curtis E. LeMay with Mackinlay Kantor, *Mission with LeMay*, p. 98.
27. For equipment deficiencies and calibration, Hq. 1st Wing, March Field, to CO, Maxwell Field, 22 Apr. 1935, subject: Avigation training and instrument flying, pp. 4–5; for the quotation, Training Directive AN–4, Air Corps Tactical School, 1934–1935, both in AFArc 248.2016J. Copies of the paper computer are in AFArc 248.2017J. One of the textbooks, 2d Lt. Thomas L. Thurlow, *Celestial Avigation*, is in USAF Academy Library Special Collections.
28. 5th Bomb Group, Hickham Field, Training directive, 1 July 1938–30 June 1939, AFArc GP–5–SU–DI(BOMB); Baldwin, Individual Training of Navigators in the AAF, p. 60. For the quotations, Daillière, "Le point astronomique dans la navigation aérienne," p. 531; and Capt. Grandison Gardner, Maxwell Field, to Ed Langmead, Wright Field, 15 Sep. 1934, AFArc 248.2016J.
29. Hq. 1st Wing, March Field, to CO, Maxwell Field, 22 Apr. 1935, subject: Avigation training and instrument flying, p. 7, AFArc 248.2016J.
30. U.S., War Dept., Air Corps Tactical School, *Bombardment Aviation Course* (Feb. 1931), p. 8, AFArc 248.101–9; History of Langley Field, 1 Mar. 1935–7 Dec. 1941, 1: 40, AFArc 285.49–2.
31. *Air Corps News Letter* 21 (1 Mar. 1938): 4.
32. 19th Bomb Gp., March Field, Operations Memo #3, 30 Aug. 1938, AFArc GP–19–SU, 1937–1938, vol. 1.
33. CG, GHQ AF, to CG, 2d Wing, Langley Field, 28 May 1936, subject: Navigation units of bombardment groups, AFArc 285.49–2, vol. 2.
34. A copy is folded into the bombardment textbook in AFArc 248.101–9.
35. Memos from Chief, Plans Section, to Chief of the Air Corps, Study on aeronautical ratings, 13 Dec. 1937, and Procurement of additional officers for the Air Corps for duty as navigators and bombers, 20 Jan. 1938, both in AFArc 145.91–174; Air Corps Circular 50–10, 30 June 1941, copy in Harbold Collection, Air University Library;

Chief of the Air Corps to Adjutant General, 12 Aug. 1941, subject: Commissioning of cadets on completing training as bombardiers and navigators, AFArc 145.91–174.
36. CG, GHQ AF, to CG, 2d Wing, Langley Field, 9 July 1935, subject: Navigation training, AFArc 285.49–2, vol. 2. A student in one of the last navigation courses at March had recommended such a step (Hq., 1st Wing, March Field, to CO, Maxwell Field, 22 Apr. 1935, subject: Avigation training and instrument flying, AFArc 248.2016J), but I am unable to connect the recommendation with the action taken.
37. CG, GHQ AF, to CG, 1st Wing, March Field, 3 July 1935, subject: Celestial navigation, AFArc 145.91–417.
38. Gen. Arnold's 1st indorsement to above, 22 July 1935, ibid.
39. Thurlow to Weems, 7 Dec. 1935, WRes; Baldwin, Individual Training of Navigators in the AAF, p. 56.
40. LeMay and Kantor, *Mission with LeMay,* p. 114; *Air Corps News Letter* 21 (1 Mar. 1938): 6 for the quotation.
41. Commander, 2d Bomb Group, Langley Field, to CG, 2d Wing, 19 Oct. 1935, subject: Navigation training, AFArc 285.49–2, vol. 2.
42. 1st indorsement, 20 July 1941, by Chief of the Air Corps, to letter of unknown origin, subject: Navigation training, AFArc 145.91–416; History of Army Air Forces Eastern Flying Training Command, 1 Jan. 1939–7 Dec. 1941, 1: 490 for the quotation, AFArc 222.01.
43. Copy of memo by Maj. Gen. Foulois, 30 Aug. 1935; CG, GHQ AF, to Adjutant General, 30 Apr. 1936, subject: Stations for reconnaissance-bombardment units; Proceedings of a board of officers appointed . . . 5 May 1939 . . . for the purpose of investigating . . . the commercial usage of the Link Training Device . . . ; Memo by Lt. Col. Carl Spaatz, Plans Div., to Chief of the Air Corps, 20 Sep. 1939, subject: Postgraduate school of flying; all in AFArc 145.91–417; History of Langley Field, 1 Mar. 1935–7 Dec. 1941, 1: 55–56, AFArc 285.49–2.
44. Capt. James E. Parker to Chief of Staff, through channels, 24 Jan. 1936, cited in Baldwin, Individual Training of Navigators in the AAF, p. 102; ibid., pp. 103–04; for the quotation, Statement of Charles J. Lunn, Director of the AAF Pan-American Airways Navigation Section and the navigator on the Lisbon flight, in History of the Army Air Forces Training Detachment, Coral Gables, 1 Aug. 1940–1 Mar. 1944, AFArc 234.667–1, vol. 1; History of Army Air Forces Eastern Flying Training Command, 1 Jan. 1939–7 Dec. 1941, 1: 491, AFArc 222.01.
45. History of Army Air Forces Eastern Fying Training Command, 1 Jan. 1939–7 Dec. 1941, 1: 493–505, quotation is on p. 499, in AFArc 222.01; Statement by Capt. Roger H. Ter-

zian, a member of the first class, in History of the Army Air Forces Training Detachment, Pan-American Airways, Coral Gables, 1 Aug. 1940–1 Mar. 1944, AFArc 234.667–1, vol. 1.

46. History of Army Air Forces Eastern Flying Training Command, 1 Jan. 1939–7 Dec. 1941, 1: 504, AFArc 222.01; Statement by Charles J. Lunn, Director of the AAF Pan-American Airways Navigation Section, in History of the Army Air Forces Training Detachment, Coral Gables, 1 Aug. 1940–1 Mar. 1944, AFArc 234.667–1, vol. 1.

47. History of Army Air Forces Training Command, 1 Jan. 1939–VJ Day, Barksdale Field, La., 15 June 1946, 4: 803–40, AFArc 220.01; Navigation training schedule, no source given, 9 June 1941, AFArc 145.91–434; Baldwin, Individual Training of Navigators in the AAF, pp. 8–9, 74–75, 89, 162–63; Program of Instruction, Training of aerial navigators for military students to be given in Air Corps Flying School, no source given, 15 July 1941, AFArc 101–27; Harbold, *Log of Air Navigation,* pp. 84–103.

48. Chief of the Air Corps to Adjutant General, 2 July 1940, subject: Procurement of trainees for non-pilot members of combat crews, AFArc 145.91–174; History of Army Air Forces Training Command, 1 Jan. 1939–VJ Day, 4: 838–41, AFArc 220.01. For the aptitude scores, Baldwin, Individual Training of Navigators in the AAF, p. 34.

49. Highlights of the Langley Field–Panama Flight, Feb. 4–12, 1937, typescript, AFArc 285.49–2, vol. 2; HQ GHQ AF, Memorandum #5, Personnel and Training, 6 Mar. 1937, AFArc 145.91–222; LeMay and Kantor, *Mission with LeMay,* pp. 153–63, 177–79; *Air Corps News Letter* 21 (1 Mar. 1938): 1–2, (15 Aug. 1938): 1–3; 22 (1 Mar. 1939): 2–5; Maj. V. J. Meloy to CO, 2d Bomb Group, Langley Field, 18 Aug. 1938, AFArc 285.49–2, vol. 2.

50. Weems to Waghorn, 2 May 1938, WML, Astrograph.

51. LeMay and Kantor, *Mission with LeMay,* pp. 142–51, 183–93; *Air Corps News Letter* 21 (15 June 1938): 3, (1 July 1938): 6; 22 (1 Apr. 1939): 6–7.

52. Maj. Gen. Oscar Westover, "1936 Passes in Review; the Army," p. 64 for the quotation; 19th Bomb Group to CG, 1st Wing, March Field, 8 Feb. 1939, subject: Analysis of bombing training, 19th Bomb Gp., 1937 and 1938, AFArc GP–19–SU, 1937–1938, vol. 1.

53. 19th Bomb Group to CG, 1st Wing, March Field, 8 Feb. 1939, subject: Analysis of bombing training, 19th Bomb Gp., 1937 and 1938, AFArc GP–19–SU, 1937–1938, vol. 1; Harris B. Hull, "The GHQ Air Force Maneuvers," p. 10.

54. Griffiths to Weems, 27 Jan. 1940, WML, Astrograph.

55. James Phinney Baxter, 3d, *Scientists against Time,* pp. 139–

40; Lt. Cmdr. Fletcher G. Watson and Henrietta H. Swope, "Loran; the Planning, Growth, and Use of Loran," pp. 11–12.
56. V. I. Zelov, commenting on paper by George R. Fairlamb, in *Aeronautical Engineering* 1 (July–Sep. 1929): 156 for the quotation; Christopher W. C. Wheatley, letter to the editor, *Flight* 6 (13 Nov. 1914): 1127–28; Gatty, "Aerial Navigation," p. 154; Byrd, *Skyward,* p. 356; Franklin L. Hunt, "Aeronautic Instruments," pp. 761–62; Werner Immler, "Die dynamische Auffassung navigatorischer Probleme."
57. Capt. F. Jansen, University of Southern California, to Henry Hughes & Son, 22 Oct. 1936, WRes.
58. Griffiths to Weems, 2 May 1938, WML, Astrograph.
59. Vannevar Bush, Foreword to Baxter, *Scientists against Time,* p. vi; Rowe, *One Story of Radar,* pp. 2, 46.
60. Memo of the Asst. Chief of Staff G–4 to the Chief of Staff, 10 Mar. 1937, subject: War Department research and development program, FY 1939, cited in Mark Skinner Watson, *Chief of Staff; Prewar Plans and Preparations,* p. 43; Gen. Henry H. Arnold, "Science and Air Power," p. 193.
61. Bertrand de Jouvenel, *The Art of Conjecture,* p. 150.

BIBLIOGRAPHY OF PRINCIPAL SOURCES

BIBLIOGRAPHIES

Ristow, Walter W. *Aviation Cartography; a Historico-Bibliographic Study of Aeronautical Charts.* 2d ed. Washington, 1960. Contains digests of and quotations from early articles not easily obtained elsewhere. Unreliable in spelling foreign words.

Smithsonian Institution. *Bibliography of Aeronautics,* by Paul Brockett. "Smithsonian Miscellaneous Collections," vol. 55. Washington, 1910. Detailed coverage of world literature to July 1909.

U.S. National Advisory Committee for Aeronautics. *Bibliography of Aeronautics,* 1909–32, in 14 vols., by Paul Brockett. GPO, 1921–36. Continuation of previous entry.

———. Report #125, "General Classification of Instruments and Problems Including Bibliography," by Mayo D. Hersey. In NACA, *7th Annual Report, 1921,* pp. 489–98. Includes wartime work.

U.S. Library of Congress. Division of Aeronautics. *The Aeronautical Index for 1939; a Subject and Author Index to Aeronautical Periodicals and Technical Reports.* New York, 1943.

———. *Subject Index to Aeronautical Periodical Literature and Reports for the Year 1938.* New York: Works Projects Administration, 1939. Mimeographed.

U.S. Works Progress Administration. *Bibliography of Aeronautics.* Part 22, "Radio," 1938; Part 25, "Air Navigation," 1939; Part 26, "Flight Instruments," 1939. New York. Mimeographed. These bibliographies are neither as thorough nor as accurate as the Brockett series, cited under Smithsonian and U.S., NACA, above; but there is no better coverage for 1933–37.

MANUSCRIPT COLLECTIONS

United States Air Force Academy, Colorado, Library, Special Collections. Papers of General William Mitchell; the Gimbel Collection, recently received and only partially cataloged, contains a variety of early materials.

United States Air Force Archives, Maxwell Air Force Base, Alabama. (In the notes, AFArc.) Correspondence, textbooks, and other training materials of the Air Corps Tactical School; correspondence of the Plans Division, Office of the Chief of the Air Corps; unpublished unit histories, which frequently reproduce documents.

Weems Memorial Library, Division of Naval History, Smithsonian Institution, Washington, D.C. (In the notes, WML.) Correspondence of Captain Philip Van Horn Weems (USN, Retired) and of his company, Weems System of Navigation. This material is available on microfilm. The title of the reel is cited in the notes, but the frames were not numbered. Weems Residence, Randall House, Annapolis, Maryland. (In the notes, WRes.) Correspondence that Captain Weems had not, as of summer 1968, delivered to the Smithsonian.

PUBLISHED SOURCES CITED

Signed government documents are alphabetized under the name of the author. Others are listed under the name of the government; only the major office (ministry or department) is considered in alphabetization. Unsigned articles appear only in the notes.

Adcock, Frank. "Some Early Observations on Aircraft with the Four-Aerial Direction-Finder." *Institution of Electrical Engineers Journal* 64 (Aug. 1926): 837–38.

Alexanderson, E. F. W. "Height of Airplane above Ground by Radio Echo." In International Civil Aeronautics Conference, Washington, 1928, *Papers Submitted,* pp. 615–18. GPO, 1928.

Alt, Eugen. "Astronomisches Ortsbestimmung im Luftschiff." *Deutsche Luftfahrer-Zeitschrift* 16 (20 Mar., 3 Apr. 1912): 131–32, 166–69.

Amundsen, Roald Engelbregt, and Lincoln Ellsworth. *First Crossing of the Polar Sea.* New York, 1927.

Aquino, Francisco Radler de. *Altitude and Azimuth Tables for Facilitating the Determination of Lines of Position and Geographical Position at Sea, the Simplest and Readiest in Solution.* Annapolis, 1908. Reprinted from *U. S. Naval Institute Proceedings,* vol. 34.

Ardouin-Dumazet. "Démonstration élémentaire de propriétés des cartes orthodromiques conformes de M. Kahn." *Revue de L'armée de l'air* 6 (Dec. 1934): 1417–29.

Arendt. "Ueber die Bedeutung magnetischer Beobachtungen im Ballon; nach einem Vortrage des Herrn Professor Dr. Eschenhagen." *Zeitschrift für Luftschiffahrt und Physik der Atmosphäre* 17 (Heft 9–10, 1898): 205–10.

Armstrong, William. *Pioneer Pilot.* London, 1952.

Arnold, Henry Harley. *Airmen and Aircraft; an Introduction to Aeronautics.* New York, 1926.

———. "Science and Air Power." *Air Affairs; an International Journal* 1 (Dec. 1946): 193.

——— and Ira Clarence Eaker. *Winged Warfare.* New York, 1941.

Ault, James Percy. "Navigation of Aircraft by Astronomical Methods." *Carnegie Institution of Washington Publications* 5: 315–37. Washington, 1926.

Aussenac, L. "Les cinémodérivomètres modernes." *Aéronautique* 16 (Apr. 1934): 77–82.

Bacon, John M. "Steering Balloons

by Upper Air Currents." *Everybody's Magazine* 5 (Aug. 1901): 171–76.

Baden-Powell, Baden Fletcher Smyth. *Ballooning as a Sport.* Edinburgh, 1907.

[Bahn, Heinrich.] "In a German Airship over England, a Story of the Raid of 19th-20th October 1917." *Royal United Service Institution Journal* 71 (Feb. 1926): 107–18.

Baidukov, George. *Over the North Pole.* Translated by Jessica Smith. New York, 1938.

Bairstow, Leonard. "Discussion of the Action of Different Types of Air-Speed Indicator." Great Britain, Advisory Committee for Aeronautics, Reports and Memoranda #145, in *Technical Report for the Year 1913–14*, pp. 461–67.

Baker, T. Thorne. "Wireless Telegraphy and Flight." *Flight* 3 (11 Mar. 1911): 198–99.

Baker, Thomas Yeomans. "The Baker Air Sextant." *Optical Society of London Transactions* 20 (June 1919): 340–42.

——— and Louis Napoleon George Filon. *Position Fixing in Aircraft during Long Distance Flights over the Sea.* London: Royal Aeronautical Society, 1920. 24 pp.

Balchen, Bernt. *Come North with Me; an Autobiography.* New York, 1958.

Baldit. "Observations sur la méthode de détermination du point en ballon." In Commission permanente internationale d'aéronautique, Extraordinary session, Brussels, Sep. 1907, *Procès verbaux et comptes rendus des travaux,* pp. 67–71. Paris, 1908.

Bamler, Karl. "Deutsche Luftfahrerkarten." *Deutsche Luftfahrer-Zeitschrift* 16 (7, 21 Feb. 1912): 59–60, 76–78.

Bardot, F. "Guidage radioélectrique des aéronefs." *Revue de L'armée de l'air* 6 (Mar. 1934): 265–82.

Bastide, A. "De quelques procédés récents de points astronomiques en vol." *Aérotechnique* 12 (Oct. 1934): 89–97.

———. Review of *Éphémérides aéronautiques pour 1937* (Paris, 1936). *Revue de L'armée de l'air,* no. 97 (Aug. 1937): 951–58.

Bastien, A. "Long Distance and Night Navigation." Translated by Francis Chichester. *Guild of Air Pilots and Air Navigators of the British Empire Journal* 6 (July 1939): 12–18.

———. "Le point astronomique." *Aérophile* 45 (May, July 1937): 111–12, 164–65; 46 (Mar. 1938): 68–69.

Bates, Mortimer F. "Filling Up the Instrument Panel." *Aviation* 40 (Aug. 1941): 90–91, 192, 194, 196, 234.

Baxter, James Phinney, III. *Scientists against Time.* Boston, 1946.

Beaumont, André [Jean Louis Camille Conneau]. *My Three Big Flights.* London, 1912.

Beij, K. Hilding. "Astronomical Methods in Aerial Navigation." U.S., NACA, Report #128. In *10th Annual Report, 1924,* pp. 271–320. GPO, 1925.

Bellini, E. "Errors of Direction Finders." *Scientific American*

Bibliography of Principal Sources

Monthly 3 (May 1921): 451–53. Reprinted from *The Electrician*, 18 Feb. 1921.

Bennett, Donald Clifford Tyndall. *Pathfinder; a War Autobiography*. London, 1958.

Benson, Charles Beverley. *Map Reading for Aviators, with a Chapter on Aerial Navigation*. New York, 1918.

Berget, Alphonse. "Mesure des hautes altitudes; méthode gravimétrique de sensibilité constante." *Aérophile* 17 (15 Feb. 1909): 90.

Bestelmeyer, A. "Zur Benutzung des Kompasses im Ballon." *Deutsche Zeitschrift für Luftschiffahrt* 14 (30 Nov. 1910): 17–18.

Bewsher, Paul. *"Green Balls"; the Adventures of a Night-Bomber*. Edinburgh, 1919.

Bidlingmaier. "Über die magnetische Ortsbestimmung im Ballon." *Illustrierte aeronautische Mitteilungen* 13 (3 Nov. 1909): 975–83.

Boeing Aircraft Company. *Handbook of Instructions for Operation of the Boeing Model 314 Flying Boat*. Seattle, 1939.

Booth, Harris. "The Theory of the Gravity-Controlled Air Speed Indicator as Applied to Dynamic Lift Flying Machines." *Aeronautical Journal* 18 (Jan. 1914): 56–58.

———. "A Warning to Pilots." *Flight* 6 (3 Jan. 1914): 4.

Bouck, Zeh. "W2XBQ Flies to Bermuda." *Radio News* 12 (July 1930): 12–13, 72–73.

Bourdin, J. "Sur un instrument analogue au compas aéronautique, décrit par M. Janssen." *Académie des sciences, Comptes rendus* 72 (6 Mar. 1871): 256–57.

Boykow, Hans. "Orientierung und Navigation im Luftfahrzeug." *Deutsche Luftfahrer-Zeitschrift* 16 (24 Oct. 1912): 530–34.

———. "Probleme der terrestrischen Navigation im Luftfahrzeuge." In Wissenschaftliche Gesellschaft für Luftfahrt, *Arbeiten zur Luftnavigierung*, pp. 32–39. Munich, 1927.

———. "Transatlantic Flight." *Scientific American Supplement* 78 (25 July 1914): 62–64. Translated from *Zeitschrift für Flugtechnik und Motorluftschiffahrt*.

Boys, C. V. "Position Finding without an Horizon." *Nature* 81 (22 July, 19 Aug. 1909): 111–13, 231–32.

Brachner, W. S. "Training in Military Aviation." *Scientific American Supplement* 83 (21 Apr. 1917): 242–43.

Brackley, Frida Helena, compiler. *Brackles; Memoirs of a Pioneer of Civil Aviation*. Norfolk, England, 1952.

Branch, William James Vivian, and E. Brook-Williams. *A Short History of Navigation*. Annapolis and Glasgow, 1942.

Brancker, Sefton. "Aerial Transport from the Business Point of View." *Flight* 12 (6 May 1920): 507.

Brett, R. Dallas. *History of British Aviation, 1908–1914*. London: Hamilton, 1934. Also in 2 vols., London: The Aviation Bookclub, n. d.

Briand, Paul L., Jr., *Daughter of the Sky; the Story of Amelia Earhart*. New York, 1960.

Briggs, Lyman J. "Aeronautical Research at the National Bureau of Standards." In International Civil Aeronautics Conference, Washington, 1928, *Papers Submitted*, pp. 483-90.

——— and others. "Present Status of Aircraft Instruments. U.S., NACA, Report # 371. In *17th Annual Report, 1931*, pp. 153-74. GPO, 1932.

Brill, Alfred. "Über eine neue einheitliche Methode zur nautischen und aeronautischen Ortsbestimmung aus Gestirnsmessungen." *Annalen der Hydrographie und maritimen Meteorologie* 38 (Nov., Dec. 1910): 614-22, 674-86.

———. "Ein Verfahren zur Auswertung astronomischer Positionsbestimmungen." *Illustrierte aeronautische Mitteilungen* 13 (3 Nov. 1909): 988-94.

Brombacher, William G. "Measurement of Altitude in Blind Flying." U.S., NACA, Technical Note #503. Washington, 1934. Mimeographed.

Brown, Arthur Whitten, assisted by Alan Bott. *Flying the Atlantic in Sixteen Hours*. New York, 1920.

Burgess, George Kimball. "Aircraft Radio Beacon Development." *Aviation* 24 (18 June 1928): 1764-65, 1798-1803.

Busignies, Henri. "Appareils indicateurs donnant par lecture directe la direction d'une onde." *Onde électrique* 6 (July 1927): 277-303.

Büttner, Alexander. "Throwing Bombs from Airships; How to Aim the Projectile to Hit a Given Object." *Scientific American Supplement* 81 (5 Feb. 1916): 85. Translated from *Die Umschau*.

Byrd, Richard Evelyn. *Little America; Aerial Exploration in the Antarctic; the Flight to the South Pole*. New York, 1930.

———. "Navigation Problems of the MacMillan Expedition." *Aero Digest* 7 (July 1925): 369, 394.

———. "Our Transatlantic Flight." *Aero Digest* 11 (July 1927): 8-10.

———. *Skyward; Man's Mastery of the Air as Shown by the Brilliant Flights of America's Leading Air Explorer*. New York, 1928.

Cailletet, Louis Paul. "Appareil destiné à mesurer les hauteurs atteintes par les aérostats." *Académie des sciences, Comptes rendus* 125 (26 Oct. 1897): 587-89.

Calahan, Harold Augustin. *The Sky and the Sailor; a History of Celestial Navigation*. New York, 1952.

Caldwell, Cy. "Getting Lost in Fog or Stormy Weather; Plea for Better Training in Instrument Flying." *Aero Digest* 26 (Feb. 1935): 27, 76-77.

Canada. Department of National Defence. *Report on Civil Aviation, 1924*. Ottawa, 1925.

Capper, J. E. "The Gordon-Bennett Cup of 1906." *Aeronautical Journal* 11 (Jan. 1907): 15-18.

Carbonara, Victor E. "Aircraft Instruments for Oversea Navigation." *Society of Automotive Engineers Journal* 21 (Nov. 1927): 528-31; 22 (June 1928): 670-73.

Bibliography of Principal Sources

Card, Stanton Freeland. *Air Navigation, Notes and Examples.* London, 1919.

Carr, C. B. "Air Route Radio Services in Great Britain." *Aviation* 27 (26 Oct. 1929): 834–38.

Cave-Browne-Cave, H. M. "Cruise of the Royal Air Force Far East Flight." *Royal Aeronautical Society Journal* 33 (July 1929): 542–72.

Chamberlin, Clarence Duncan. *Record Flights.* Philadelphia, 1928.

Chichester, Francis Charles. "Navigation—Fourth Rate or First Class?" *Flight* 35 (19 Jan. 1939): 54–56.

———. *Ride on the Wind.* New York, 1937.

———. *Seaplane Solo.* New York, 1934.

Clarke, Nell Ray. "Are We over the Pole?" *Scientific American* 135 (Sep. 1926): 188–89.

Clemence, G. M., and D. H. Sadler. "Unification of the Air Almanac and the American Air Almanac." *Navigation, Journal of the Institute of Navigation* 3 (Sep. 1951): 9–11.

Clift, Eric Hollocoombe. Letter to the Editor. *Flight* 3 (13 May 1911): 429.

———. "The Magnetic Compass; Its Construction and Use." *Flight* 4 (24 Feb., 2 Mar. 1912): 163–65, 199–201.

C. O. [Creagh-Osborne ?]. Letter to the Editor. *Flight* 3 (6 May 1911): 409.

Collinder, Per Arne. *History of Marine Navigation.* Translated by Maurice Michael. London, 1954.

Collins, F. A. "Land Lighthouses." *Aeronautical Digest* 1 (Nov. 1922): 196.

Congrès international d'aéronautique, IVe, Nancy, 1909. *Procès-verbaux; rapports & mémoires.* Paris, 1909.

Cooper, Bertram G. "Aeronautical Maps and Signs." *Aeronautical Journal* 15 (Oct. 1911): 142–45.

———. "Air-Speed Indicators," an editorial. Ibid. 18 (Jan. 1914): 55.

———. "Meeting of the International Commission for Aeronautical Maps." Ibid. 15 (July 1911): 123–24.

Cordeiro, F. J. B. "Pulfrich Sextant." *Scientific American Monthly* 1 (Jan. 1920): 44.

Corrigan, Douglas. *That's My Story.* New York, 1938.

Courtney, Frank T. "Future of Air Navigation." *Aviation* 29 (Nov. 1930): 264–68.

Coutinho, Gago, and Jorge de Castilho. "Navigation des avions du service commercial pour l'Amérique du Sud." In IV Congresso internazionale di navigazione aerea, Rome, 1927, *Memorie* 2: 100–25.

Craddock, Christopher. *Whispers from the Fleet.* Portsmouth, England, 1907.

Creagh-Osborne, Frank. *The Magnetic Compass in Aircraft.* 3d ed. London, 1915.

Curtiss, R. H. "Rise of Navigation." *Scientific American Supplement* 86 (14 Dec. 1918): 370–71, 378. From *Popular Astronomy.*

Cutrell, E. A. "Instrument and Radio Flying." *Aviation* 34 (June 1935): 11–14.

Daillière. "Le point astronomique

dans la navigation aérienne." *Revue des forces aériennes* 5 (Apr., May 1933): 391–411, 511–34.

Dainty, W. Reginald. "Radio-Telegraphic Apparatus for Aircraft." *Flight* 6 (25 Dec. 1914): 1230–31.

Daloz, G. "La boussole aérienne." *Revue aérienne* 3 (25 Nov. 1910): 652–58. Similar article in *Aérophile* 18 (1 Dec. 1910): 531–33.

Daniel Guggenheim Fund for the Promotion of Aeronautics. *Equipment Used in Experiments to Solve the Problem of Fog Flying; a Record of the Instruments and Experience of the Fund's Full Flight Laboratory.* New York, 1930.

Darwin, Horace. "Scientific Instruments, Their Design and Use in Aeronautics," the 1st Wilbur Wright Memorial Lecture read before the Aeronautical Society of Great Britain, 21 May 1913. *Flight* 5 (31 May 1913): 599–600.

———. "Scientific Instruments Used in Aeronautics." *Engineering Magazine* 45 (Sep. 1913): 878–80.

———. "The Static Head Turn Indicator for Aeroplanes." *Flight* 11 (30 Oct. 1919): 1429–31.

Davies, L. Graham. "Steering by Compass." *Flight* 2 (24 Sep. 1910): 775–76.

Davies, R. E. G. *A History of the World's Airlines.* London, 1964.

De Havilland, Geoffrey. *Sky Fever; the Autobiography of Sir Geoffrey de Havilland, C. B. E.* London, 1961.

Delens, Paul. "La dérive dans le vent." *Aérophile* 18 (1 May 1910): 203–04.

———. "Étude des parcours aériens." Ibid. (15 June 1910): 275–76.

Dellinger, J. H., and Haraden Pratt. "Development of Radio Aids to Air Navigation." *Institute of Radio Engineers Proceedings* 16 (July 1928): 890–920.

Delsasso, Leo P. "The Acoustic Altimeter." *Science* 85 (28 May 1937, supplement): 10, 12.

Deniélou, M. "T. S. F. et navigation aérienne." *Aéronautique* 2 (Nov. 1920): 233–38.

Denman, Roderick. "Radio Air Navigation; the Trend of British Development." *Flight* 31 (21 Jan. 1937): 54–56.

Dex, Léo. "La détermination du point en ballon." *Revue scientifique* 50 (10 Dec. 1892): 743–50.

Dickson, Bertram. "A Safety Speed-Alarm for Aeroplanes." *Flight* 2 (20 Aug. 1910): 658.

Dienstbach, Carl. "The Gyrotelescope." *Scientific American* 113 (23 Oct. 1915): 363, 368.

Dixie, Albert Edward. *Air Navigation for Flight Officers.* 2d ed. London, 1918.

Dixon, Charles. *The Conquest of the Atlantic by Air.* London, 1931.

Dobson, Gordon M. B. "Design of Instruments for Navigation of Aircraft." *The Geographical Journal* 56 (Nov. 1920): 370–89.

Douhet, Giulio. *The Command of the Air.* Translated by Dino Ferrari from the 2d ed., 1927. New York, 1942.

Draper, C. S. "Sonic Altimeter for

Aircraft." U.S., NACA, Technical Note #611. Washington, 1937. Mimeographed.

Driencourt, L. "Le planisphère de Mercator et la navigation maritime ou aérienne par relèvements radiogoniométriques pris du bord." *Revue aéronautique internationale* 6 (Mar. 1936): 75–79.

Drodofsky, M. "Astronomische Beobachtungsinstrumente." In Germany, Reichsluftfahrtministerium, *Ringbuch der Luftfahrttechnik*, section VB2, 30 Sep. 1937.

Dubilier, William. "Wireless Communication and Aeronautics." *Scientific American Supplement* 78 (19 Sep. 1914): 189–90.

Duke, Neville, and Edward Lanchbery, eds. *The Saga of Flight; an Anthology*. New York, 1961.

Duncan, Richard. *Air Navigation and Meteorology*. 3d ed. Chicago, 1936.

Dunmore, F. W. "Radio Beacons Non-Directive and Directive." *Radio News* 5 (May 1924): 1554–56, 1685–86.

Dunoyer, L. "Le point estimé en navigation aérienne." *Aéronautique* 1 (Sep. 1919): 133–45.

Dürr, Ludwig. *Fünfundzwanzig Jahre Zeppelin-Luftschiffbau*. Berlin, 1924.

Duval, Albert B. "Le cercle calculateur de routes et de vitesses." *Aéronautique* 2 (Sep. 1920): 155–57.

———. "L'emploi des méthodes de navigation est indispensable au succès des transports aériens commerciaux." In International Congress of Aerial Navigation, 2d, London, 1923, *Report*, pp. 765–68.

———. "Maps and Navigation Methods." *Aerial Age* 15 (8 May 1922): 198–99.

——— and Léo Hébrard. *Traité pratique de navigation aérienne*. Paris, 1922. Previously appeared serially in *Aéronautique*.

Earhart, Amelia. *Last Flight*. Arranged by George Palmer Putnam. New York, 1937.

Eaton, Herbert Nelson. "Aerial Navigation." *U.S. Air Service* 8 (Sep., Oct., Dec. 1923): 37–40, 39–44, 45–47.

———. "Aerial Navigation and Navigating Instruments." U.S., NACA, Report #131. In *7th Annual Report, 1921*, pp. 762–802. GPO, 1923.

Eberle, William Caveny, and P. V. H. Weems. *Learning to Navigate*. New York, 1939.

Eckener, Hugo. *My Zeppelins*. Translated and abridged by Douglas Robinson. London, 1958.

Edgerton, James C. "Radio as Applied to Air Navigation in the Air Mail Service." *U.S. Air Service* 5 (Feb. 1921): 12–14.

Ellington, E. L. "The Present Position of Aircraft Research and Contemplated Developments." In Great Britain, Air Ministry, "Proceedings of the Air Conference," Cmd. 1157, *Parliamentary Papers*, 1921, 8: 57–65.

Engel, Leonard H. "Principle of the Echo Altimeter." *Science* 88 (11 Nov. 1938, supplement): 8, 10.

Erskine-Murray, J. "Determining Position at Sea by Wireless."

Engineering Magazine 44 (Mar. 1913): 949–51.

Espenschied, Lloyd, and R. C. Newhouse. "A Terrain Clearance Indicator." *Bell System Technical Journal* 18 (Jan. 1939): 222–34.

Eyb, Robert. *Fliegerhandbuch; ein Leitfaden der gesamten Flugtechnik.* Vol. 13 in the Bibliothek für Luftschiffahrt und Flugtechnik. 2d ed. Berlin, 1916.

Fairlamb, George R., Jr. "Air Navigation; an Estimate of Progress Made, and a Discussion of the Minimum of Theoretical Knowledge Required by Pilots." *Aeronautical Engineering* 1 (July–Sep. 1929): 151–57.

Fales, Elisha Noel. *Learning to Fly in the US Army; a Manual of Aviation Practice.* New York, 1917.

Favé, L. "Recherches sur les instruments et les méthodes propres à la détermination du point en ballon." In IIIe congrès international d'aéronautique, Milan, 1906, *Rapports & mémoires,* pp. 145–79. Paris, 1907.

Felix, Edgar H. "Aircraft Radio Direction Finding Equipment Developed by the Navy Department." *Aerial Age Weekly* 9 (14 July 1919): 852–53.

Fink, Don. "Aviation Radio," a column. *Aviation* 37 (Nov. 1938): 48–49, 80.

———. "Coast Guard Radio." Ibid. (May 1938): 48–49, 70.

Finney, Robert T. *History of the Air Corps Tactical School, 1920–1940.* USAF Historical Studies no. 100. Maxwell Air Force Base, Ala., 1955.

Fischer, Karl T. "Ein neues Barometer ('Luftdruckaräometer')." *Illustrierte aeronautische Mitteilungen* 4 (July 1900): 68–70.

Fischer, P. B. "Astronomische Ortsbestimmung im Ballon nach Schwarzschild." *Deutsche Zeitschrift für Luftschiffahrt* 14 (9 Feb. 1910): 22–23.

Fitch, Willis Stetson. *Wings in the Night.* Boston, 1938.

Fleming, H. C. "How the ZR 3 Was Piloted by Radio." *Radio News* 6 (Jan. 1925): 1134–35, 1291–92, 1294.

Fonvielle, Wilfrid de. "Nouvelle méthode altimétrique applicable en ballon." *Aérophile* 17 (1 Feb. 1909): 65.

Förstner, Gustav. "Genauigkeit von Höhenbeobachtungen mit dem Periskopsextanten." Deutsche Versuchsanstalt für Luftfahrt, report #349, abstract in *Aero Digest* 25 (July 1934): 29. Text of DVL report #356, same title, in *Zeitschrift für Flugtechnik und Motorluftschiffahrt* 24 (28 Dec. 1933): 680–84.

Franck, P. "Le radiocompas et la navigation aérienne." *Onde électrique* 7 (Mar. 1928): 109–18.

Frankenberg. "Kirchturmuhren, ein Orientierungsmittel für Luftfahrzeuge." *Deutsche Zeitschrift für Luftschiffahrt* 15 (19 Apr. 1911): 19.

———. "Terrestrisches Karten-Orientierungs-System." *Deutsche Luftfahrer-Zeitschrift* 16 (21 Feb. 1912): 80–82.

Fredette, Raymond H. *The Sky on Fire; the First Battle of Britain, 1917–1918, and the Birth of the*

Royal Air Force. New York, 1966.

Friedensburg, Walter. "The Distant Compass." *Aerial Age Weekly* 13 (29 Aug. 1921): 587-89.

Gamble, Charles Frederick Snowden. *The Story of a North Sea Air Station; Being Some Account of the Early Days of the Royal Flying Corps (Naval Wing) and of the Part Played Thereafter by the Air Station at Great Yarmouth and Its Opponents during the War 1914-1918.* Oxford, 1928.

Gansberg, Fritz. *Der Flugzeugkompass und seine Handhabung; Kompasskompensieren — Kursabsetzen; ein Handbuch für Flugzeugführer und Beobachter.* Berlin, 1915.

Gasser, Max. *Eine Flugkartenstudie. Vortrag gehalten auf dem XVII. deutschen Geographentage zu Lübeck, Pfingsten 1909.* Würzburg, 1909.

———. "Ueber die Luftschiffer-Kartenfrage." *Deutsche Zeitschrift für Luftschiffahrt* 15 (13 Dec. 1911): 11-12.

Gatty, Harold. "Aerial Navigation; Methods and Equipment." *Society of Automotive Engineers Journal* 30 (Apr. 1932): 153-58, 170.

Gaty, John P. "Radio Compass; Description of Apparatus and Its Operation." *Aviation* 35 (May, June 1936): 24-26, 15-18.

Gideon, Samuel Edward. *Map Reading.* U.S. Army School of Military Aeronautics, University of Texas, Austin, 1918.

Gill, Napier John. *The Flyer's Guide; an Elementary Handbook for Aviators.* London, 1917.

Gingrich, John Edward. *Aerial and Marine Navigation Tables.* New York, 1931.

Goldsborough, Brice. "The Earth Inductor Compass." *Aero Digest* 10 (June 1927): 542, 544.

Gradenwitz, Alfred. "Lighthouses for the Aerial Navigator; Guiding the Airman at Night." *Scientific American* 110 (16 May 1914): 411-12.

Grahame-White, Claude, and Harry Harper. *Our First Airways; Their Organisation, Equipment, and Finance.* London, 1919.

Graves, Sandham. *The Lost Diary.* Victoria, B. C., 1941.

Great Britain. Admiralty. *Manual of Navigation, 1914.* London, 1915.

Great Britain. Advisory Committee for Aeronautics. "Report for the Year 1919-20." Cmd. 1120, *Parliamentary Papers,* 1921, 8: 195-246.

Great Britain. Advisory Committee on Civil Aviation. "Report on Governmental Assistance for the Development of Civil Aviation, 20 April 1920." Cmd. 770, *Parliamentary Papers,* 1920, 9: 799-810.

Great Britain. Aeronautical Research Committee. "Report for the Year 1920-21." Cmd. 1458, *Parliamentary Papers,* 1921, 8: 247-98.

Great Britain. Air Ministry. *Air Navigation Instruments (Excluding Compasses).* Air publication #803. HMSO, 1924.

———. "Annual Report on the Progress of Civil Aviation (April 1, 1925-March 31,

1926)." Cmd. 2707, *Parliamentary Papers*, 1926, 8: 97–160.

———. "Half-Yearly Report on the Progress of Civil Aviation (October 1, 1919–March 31, 1920)." Cmd. 800, *Parliamentary Papers*, 1920, 9: 811–33.

———. "Half-Yearly Report on the Progress of Civil Aviation (October 1, 1920–March 31, 1921)." Cmd. 1342, *Parliamentary Papers*, 1921, 8: 447–98.

———. "Half-Yearly Report on the Progress of Civil Aviation (April 1–September 30, 1921)." Cmd. 1559, *Parliamentary Papers*, 1921, session 2, 1:1–66.

———. "Half-Yearly Report on the Progress of Civil Aviation (October 1, 1921–March 31, 1922)." Cmd. 1710, *Parliamentary Papers*, 1922, 11: 935–62.

———. "Proceedings of the Air Conference, 1920." Cmd. 1157, *Parliamentary Papers*, 1921, 8: 299–446.

———. "Report on the Progress of Civil Aviation (April 1–December 31, 1926)." Cmd. 2844, *Parliamentary Papers*, 1927, 7: 93–156.

———. "Synopsis of Progress of Work in the Department of Civil Aviation, May 1, 1919–October 31, 1919." Cmd. 418, *Parliamentary Papers*, 1919, 33: 607–16.

Great Britain. Committee on Education and Research in Aeronautics. "Report" dated 12 Dec. 1919. Cmd. 554, *Parliamentary Papers*, 1920, 9: 783–98.

Great Britain. Ministry of Munitions. Technical Department (Aircraft Production). "Report on the Friedrichshafen Bomber." Reprinted in *Flight* 10 (18 July 1918): 793-99.

Grooch, William Stephen. *Skyway to Asia*. New York, 1936.

Guillaume, Charles-Éduard. "Rapport succinct sur les travaux scientifiques de la C.P.I.A." In IIIe congrés international d'aéronautique, Milan, 1906, *Rapports & mémoires*, pp. 61–67. Paris, 1907.

Guyot. "Un nouvel appareil de calcul rapide du point astronomique à bord d'avion." *Aero Digest* 25 (Dec. 1934): 40, English abstract from *Aéronautique* 16 (Sep. 1934): 73–81.

Haaren, Paul J. "The Seagull Flies." *Scientific American* 120 (10 May 1919): 488–89, 498–99.

Haddow, George William, and Peter Michael Grosz. *The German Giants; the Story of the R-Planes, 1914–1919*. London, 1962.

Halsted, Arthur. "Italian Semi-Rigid Airships Observed at the Front." *Air Service News Letter* 5 (26 Feb. 1921): 1–2.

Hamel, Gustav W. Letter to the Editor. *Flight* 3 (13 May 1911): 429.

——— and Charles Cyril Turner. *Flying; Some Practical Experiences*. London, 1914.

Harbold, Norris B. *The Log of Air Navigation*. San Antonio, 1970.

Hartmann, E. "Ein Libellenquadrant in neuer Form für astronomische Ortsbestimmungen (Zahnbogen - Höhenmesser)." *Deutsche Luftfahrer-Zeitschrift* 16 (2 Oct. 1912): 495–96. Also *Annalen der Hydrographie und maritimen Meteorologie* 40 (1912): 486–88.

Hartz, Robert E. "Map Reading by Aviators." *Journal of Geography* 30 (Nov. 1931): 339–41.

Haupt-Heydemarck, Georg Wilhelm. *Double-Decker C.666*. Translated by Claud W. Sykes. London, 1931.

Hawker, Harry George, and Kenneth Mackenzie Grieve. *Our Atlantic Attempt*. London, 1919.

Heath, Mary. "The Technical Aspects of a Trans-African Flight on a Light Aeroplane." In International Civil Aeronautics Conference, Washington, 1928, *Papers Submitted*, pp. 24–31. GPO, 1928.

Hébrard, Léo. "Le bombardement de nuit et les problèmes de la navigation aérienne." *Aéronautique* 2 (Sep. 1920): 158–63.

Hegenberger, Albert F. "Avigation; Some Non-Technical Observations on the Present Status and Trend of Development of Avigation." *Maryland Academy of Sciences Journal* 1 (Oct. 1930): 212–22.

——— and Bradley Jones. "Oversea Navigation." *Society of Automotive Engineers Journal* 21 (Dec. 1927): 699–702; 22 (June 1928): 670–73.

Henke, Alfred. "Langstreckennavigation." In Germany, Reichsluftfahrtministerium, *Ringbuch der Luftfahrttechnik*, Section VB8, 26 July 1940.

Herschel, Winslow H. "The Pitot Tube and Other Anemometers for Aeroplanes." U.S., NACA, Report #2, Part 1. In *1st Annual Report, March 3, 1915–June 30, 1915*, pp. 79–100. Published as 64th Congress, 1st Session, Senate Documents, vol. 7.

Hersey, Mayo D. "General Classification of Instruments and Problems Including Bibliography." U.S., NACA, Report #125. In *7th Annual Report, 1921*, pp. 477–98. GPO, 1923.

———, F. L. Hunt, and H. N. Eaton. "The Altitude Effect on Air Speed Indicators." U.S., NACA, Report #110. In *6th Annual Report, 1920*, pp. 693–717. GPO, 1921.

Hewson, Joseph Bushby. *A History of the Practice of Navigation*. Glasgow, 1951.

Heyl, Paul R., and Lyman J. Briggs. "The Earth Inductor Compass." *U.S. Air Service* 7 (Nov., Dec. 1922): 30–33, 30–33. Reprinted from *Proceedings of the American Philosophical Society* 61 (1922).

Hill, Roderic Maxwell. *The Baghdad Air Mail*. London, 1929.

Hingsburg, F. C. "Air Navigation Facilities." *Aviation* 29 (Aug. 1930): 69–74.

Hinton, Walter. "First Trans-Atlantic Flight." *American Academy of Political and Social Science Annals* 131 (May 1927): 94–100.

Horton, C. E., and C. Crampton. "Radio Compass Developed in H. M. Signal School." *Institution of Electrical Engineers Journal* 73 (Sep. 1933): 284–94.

Hucks, B. C. "A Further Three Years' Flying Experience." *Flight* 9 (14 June 1917): 588–91.

Hughes, Arthur Joseph. *History of Air Navigation.* London, 1946.

Hull, Harris B. "The GHQ Air Force Maneuvers." *Air Corps News Letter* 21 (1 June 1938): 10. Reprinted from *Sperryscope.*

Hunsaker, J. C. "How the NC Boats Were Equipped for the Pioneer Flight." *Automotive Industries* 41 (24 July 1919): 172–76.

Hunt, Franklin L. "Aeronautic Instruments." *Optical Society of America Journal and Review of Scientific Instruments* 6 (Sep. 1922): 744–811.

———. "Aircraft Speed Instruments; Airspeed Indicators." U.S., NACA, Report #127, Part 1. In *7th Annual Report, 1921,* pp. 563–81. GPO, 1923.

———. "New Types of Aircraft Instruments." *Optical Society of America Journal and Review of Scientific Instruments* 12 (Mar. 1926): 227–69.

———. "Recent Developments and Outstanding Problems." U.S., NACA, Report #132. In *7th Annual Report, 1921,* pp. 805–14. GPO, 1923.

Hyland, Lawrence A. "True Altitude Meters." *Aviation* 25 (27 Oct. 1928): 1322–23, 1356–62.

Immler, Werner. "Die dynamische Auffassung navigatorischer Probleme." *Hansa* 74 (27 Feb., 6 Mar. 1937): 448–50, 482–84.

Jacobs, A. M. "Radio Beacon Guides Night Air Mail." *Radio News* 7 (Mar. 1926): 1277.

Jäderin, Edv. "Nivåsextant, konstruerad för Andrées Polarballong." *Kongl. Vetenskaps-Akademiens Förhandlingar,* no. 9 (1897), pp. 493–505.

Jameson, P. R. "The Development of the Tycos Altimeter." *Aerial Age Weekly* 9 (17 Mar. 1919): 64–65.

Johnston, S. Paul. "Clipper Gear; Accessories, Instruments, Gadgets Aid Pan American's Pacific Project." *Aviation* 34 (Sep. 1935): 14–16.

Jones, B. Melville. "Flying over Clouds in Relation to Commercial Aeronautics." *Flight* 12 (11 Mar. 1920): 298–99.

Jones, E. Letter to the Editor. *Flight* 2 (17 Sep. 1910): 762.

Jones, E. Lester. "Aeronautic Maps." *Aerial Age Weekly* 4 (26 Feb. 1917): 693.

Jones, Henry Albert. *The War in the Air.* See Raleigh, Walter Alexander.

Josephson, Matthew. *Empire of the Air; Juan Trippe and the Struggle for World Airways.* New York, 1944.

Jouvenel, Bertrand de. *The Art of Conjecture.* Translated by Nikita Lary. New York, 1967.

Jurmont, A. "L'instinct de l'orientation chez les aviateurs." *Revue aérienne* 5 (10 Apr. 1912): 193–94.

Kahn, Louis. "Les méthodes de navigation employées par les aviateurs Costes et Bellonte, au cours de la traversée de l'Atlantique." *Génie civil* 97 (8 Nov. 1930): 469–70.

Kaiser, F. W. "Navigation für Luftschiffe." *Die Luftflotte* 1 (Mar. 1909): 1–3.

Kenly, W. L. "Report of Year Ending 30 June 1918." *Air Service Weekly News Letter* 1 (14 Dec. 1918): 8.

Kiel, Georg. "Measurement of the True Dynamic and Static Pressures in Flight." U.S., NACA, Technical memorandum #913, Washington, 1939. Mimeographed. Translated from *Luftfahrtforschung* 15 (10 Dec. 1938): 583–97.

Kimball, James H. "Problem of Transatlantic Weather Forecasting." *SAE Journal* 22 (June 1928): 670–71.

Kingsford, A. R. *Night Raiders of the Air; Being the Experiences of a Night Flying Pilot, Who Raided Hunland on Many Dark Nights during the War.* London, 1930.

Kingsford-Smith, Charles Edward, and Charles T. P. Ulm. *The Flight of the Southern Cross.* New York, 1929.

Koehl, Hermann, James C. Fitzmaurice, and Guenther von Huenefeld. *The Three Musketeers of the Air; Their Conquest of the Atlantic from East to West.* Sections by Koehl and Huenefeld translated by George F. Dunay. New York, 1928.

Kölle, Waldemar. "Marineluftschiffe im Kriege, in Sturm und Not." *Deutsche Rundschau* (Jan. 1926), pp. 49–61.

Koppe, Heinrich. *Luftnavigierung und die Arbeiten des Navigierungs-Ausschusses der WGL.* Munich, 1929, 19 pp. Reprinted from *Wissenschaftliche Gesellshaft für Luftfahrt Jahrbuch, 1929.*

Kritzinger, H. H. "Die Uhr als Orientierungsmittel im Ballon." *Deutsche Zeitschrift für Luftschiffahrt* 15 (22 Feb. 1911): 23–25.

la Baume-Pluvinel, A. de. "La détermination du point en ballon." *Aérophile* 12 (Jan. 1904): 20–22. Translated as "Ortsbestimmung im Ballon." *Wiener Luftschiffer Zeitung* 4 (Jan. 1905): 11–13.

———. "La détermination du point en ballon; rapport sur l'état actuel de la question." In IVe congrès international d'aéronautique, Nancy, 1909, *Procès-verbaux; rapports & mémoires,* pp. 82–88. Paris, 1909.

———. "Note sur l'emploi du sextant à niveau pour faire le point en ballon." In Commission permanente internationale d'aéronautique, *Procès verbaux et comptes rendus des travaux de la session extraordinaire tenue à Bruxelles du 12 au 15 septembre 1907,* pp. 63–67. Paris, 1908.

———. "Recherches sur la détermination du 'Point,' et expériences de télégraphie sans fil en ballon." *Aérophile* 16 (15 Dec. 1908): 504–05.

Lachenbruch, Jerome. "The Wireless Compass; Piloting Ships into Port by Radio." *Scientific American* 120 (22 Mar. 1919): 291, 306–07.

Ladd, Story B. "Aeronautic Roads." *Aerial Age Weekly* 13 (22 Aug. 1921): 563–64.

Langewiesche, Wolfgang. *I'll Take the High Road.* New York, 1939.

la Rougery, Blondel. "Documentation sur les instruments de navigation." *Aérophile* 44 (Apr.–Aug. 1936): 89–91, 111–12, 134–35, 159–60, 181–85.

Lees, W. Thomson. "How Radio

Serves Aviation Abroad." *Radio News* 11 (July 1929): 42–45.

Lefrance, Jean-Abel. "Dropping Bombs from Aeroplanes." *Scientific American Supplement* 84 (18 Aug. 1917): 108–09.

———. "The German Goerz Bombing Sight." *Flight* 9 (13 Sep. 1917): 939–41.

Léglise, Pierre. "Notes techniques sur une série de performances d'hydraviation marchande; records et vols du 'Lieutenant-de-Vaisseau-Paris.'" *Aéronautique* 20 (Jan. 1938): 7.

Leick, Walter. *Astronomische Ortsbestimmungen mit besonder Berücksichtigung der Luftschiffahrt.* Leipzig, 1912.

———. "Praxis der Ortsbestimmung im Ballon." *Illustrierte aeronautische Mitteilungen* 13 (3 Nov. 1909): 994–95.

LeMay, Curtis E., with Mackinlay Kantor. *Mission with LeMay; My Story.* New York, 1965.

Le Prieur, Yves. "Avec la mission de Goys, de Paris à Gao." *Académie des sciences, Comptes rendus* 180 (20 Apr. 1925): 1193–94.

———. "Le correcteur de route; nouvelle méthode de navigation aérienne à l'estime." *Aéronautique* 2 (Nov. 1920): 227–32.

Lindbergh, Anne Morrow. *Listen! the Wind.* New York, 1938.

Lindbergh, Charles Augustus. "My Flight to Paris." *Aero Digest* 10 (June 1927): 514, 516, 518, 520, 522, 524.

———. *The Spirit of St. Louis.* New York, 1953.

———. *"We"; the Famous Flier's Own Story of His Life and His Transatlantic Flight, Together with His Views on the Future of Aviation.* New York, 1927.

Lindt, W. "Ein Libellenquadrant in neuer Form für astronomische Ortsbestimmungen." *Deutsche Luftfahrer-Zeitschrift* 16 (29 May 1912): 273–74.

Long, Sidney Herbert. *Navigational Wireless.* London, 1927.

Loring, E. J. "Bombing and Bombing Sights." *Scientific American* 126 (Jan. 1922): 49–51.

Löwe, Karl F. *Flugzeugortung.* Vol. 23 in "Flugzeugbau und Luftfahrt," edited by E. Pfister. Berlin, 1934.

Lucas, Keith. "Report on the Errors of Compasses on Aeroplanes." Great Britain, Aeronautical Research Committee, Reports and memoranda #238. In *Report for 1915–16,* pp. 554–67.

Luck, David G. C. "Omnidirectional Radio Range, a New Navigational Instrument." *Aviation* 40 (Mar. 1941): 134. Abstract.

Ludewig, Paul. *Die drahtlose Telegraphie im Dienste der Luftfahrt.* Berlin, 1914.

Lutz, Frederick W. "Kruesi Radio Compass for Commercial Use." *Aero Digest* 26 (Feb. 1935): 50, 75.

Lux, Fritz. "Wellentelegraphische Ortsbestimmung für die Luftschiffahrt." *Illustrierte aeronautische Mitteilungen* 13 (3 Nov. 1909): 995–98.

McConnell, Burt M. "Aeroplane in Arctic Exploration," *Scientific American* 115 (30 Sep. 1916): 295, 308.

MacCoull, Neil. "The Sperry Drift Indicator." *Aerial Age Weekly* 1 (16 Aug. 1915): 523.

McIntosh, C. H. "Navigation with the D/F Loop." *Aviation* 40 (Sep. 1941): 52–53, 158, 160, 162; (Oct. 1941): 54–55, 148–50; (Nov. 1941): 72–73.

Macmillan, Norman, ed. *Let Experts Tell You; How We Fly.* London: Guild of Air Pilots and Air Navigators of the British Empire, 1939.

Maitland, E. M. "Log of the Pioneer Dirigible Trip." *Automotive Industries* 41 (10 July 1919): 53–58, 79–81.

———. "The Transatlantic Voyage of R 34." *Flight* 11 (10 July 1919): 906–12.

Maitland, Lester J. "Markings for American Airways." *Aeronautical Digest* 2 (May 1923): 327–29, 381.

Malevé, L. "Détermination de l'altitude en aviation." *Aérostation* 7 (1 Jan. 1910): 5–11.

Manton, Marcus Dyce. "Wireless Equipment on Aircraft." *Scientific American Supplement* 78 (3 Oct. 1914): 222.

Marcillac, P. "Les appareils pour ascensions maritimes." *Aéronautique* 1 (no. 3, 1903): 41–42.

Marcuse, Adolf. *Astronomische Ortsbestimmung im Ballon.* Berlin, 1909.

———. "Die astronomische Ortsbestimmung im Ballon und ihre Bedeutung für die Luftschiffahrt." *Illustrierte aeronautische Mitteilungen* 11 (Jan. 1907): 20–25.

Martin, James V. "Across the Atlantic by Aeroplane; the Problem and Suggestions for Its Solution." *Scientific American* 106 (3 Feb. 1912): 106, 116.

Martin, William Robert. "Navigation." *Encyclopaedia Britannica*, 11th ed., 19: 284–98.

Matsuo, Sadahiro. "Altimètre à lecture directe pour l'aéronautique." *Onde électrique* 17 (July 1938): 362–71.

Mears, A. H., H. B. Hendrickson, and W. G. Brombacher. "Altitude Instruments; Altimeters and Barographs." U.S., NACA, Report #126, Part 1. In *7th Annual Report, 1921,* pp. 501–23. GPO, 1923.

Meckel, Paul A. "Ein Beitrag zur Frage der Balloninstrumente." *Illustrierte aeronautische Mitteilungen* 13 (29 Dec. 1909): 1151–52.

———. "Orientierungssystem." *Deutsche Luftfahrer-Zeitschrift* 16 (21 Feb. 1912): 82–83.

Metz, H. I. "Cone of Silence Markers Identify Exact Locations of Range Stations." *Air Commerce Bulletin* 8 (Feb. 1937): 169–75.

Millar, John H. "Radio on the Empire Routes." *Aviation* 36 (Oct. 1937): 32–33, 75.

Moedebeck, Hermann Wilhelm L. "Aeronautische Landkarten, ein Bedürfnis für Freifahrten." *Illustrierte aeronautische Mitteilungen* 10 (Sep. 1906): 299–302.

———. "Les cartes aéronautiques de la Fédération allemande." Congrès international d'aéronautique, IVe, Nancy, 1909, *Procès-verbaux; rapports & mémoires,* pp. 89–96. Paris, 1909.

———. *Handbuch der Luftschiffahrt mit besonderer Berück-*

sichtichung ihrer militarischen Verwendung. Leipzig, 1886.

———. "Die Karte des deutschen Luftschiffer-Verbandes; Bearbeitet von der deutschen Kommission für Luftschifferkarten." *Deutsche Zeitschrift für Luftschiffahrt* 14 (12 Jan. 1910): 14–20.

———. *Pocket-Book of Aeronautics.* Translated from the 2d edition by W. Mansergh Varley, with additional material. London, 1907.

———. "Ueber das Landen mit Ballons." *Deutscher Verein zur Förderung der Luftschiffahrt Zeitschrift* 7 (no. 9, 1888): 272–76.

Möller, Johannes. "Die astronomische Ortsbestimmung im Luftschiff." *Illustrierte aeronautische Mitteilungen* 13 (3 Nov. 1909): 967–75.

Möller, W. "Kompassübertragungen und Kompasssteuerungen im Flugzeug." Cinquième congrès international de la navigation aérienne, The Hague, 1930, [*Rapports*], 1: 59–82. The Hague, 1931.

Montagu of Beaulieu, Lord. "The World's Air Routes and Their Regulation." *Flight* 9 (28 June 1917): 653–59.

Moriarty, H. A. "Navigation." *Encyclopaedia Britannica,* 9th ed., 17:264–77.

Murphy, William H. "The Interlocking Equisignal Radio Beacon." *Aero Digest* 11 (Aug. 1927): 172–75.

Neumann, Georg Paul, ed. *Die deutschen Luftstreitkräfte im Weltkriege.* Berlin, 1920.

Niemann, Wilhelm. "Langstrecken-Navigation in Luftfahrzeugen über Land und See im In- und Ausland." *Hansa* 72 (29 June–13 July 1935): 1043–47, 1081–85, 1124–28.

O'Gorman, M. "Notes on Some Scientific and Technical Aspects of the Imperial Air Mail Service." In Great Britain, Civil Aviation Advisory Board, *First Report on Imperial Air Mail Services.* Cmd. 1739, *Parliamentary Papers,* 1922, 11: 963–1024.

Ozouf, René. "The Daloz Compass." *Flight* 3 (14 Jan. 1911): 38.

Peace, John D., Jr. "Fix Calculator." *Aviation* 37 (Aug. 1938): 49, 77.

Pennow, W. A. "Marking the Sky Trails." *Electric Journal* 30 (Nov. 1933): 445–49.

Percheron, Maurice. *Utilisation de la carte et de la boussole en aviation.* Paris, 1917. 27 pp.

Pierce, Samuel S. "The Use of the Compass in Aviation." *Aerial Age Weekly* 1 (19 Apr. 1915): 111, 116.

Piumatti, Claudio. "Aeroplanes in Naval Warfare." *Flight* 3 (18 Feb. 1911): 130–32. Translated from *Rivista marittima* by R. H. Keate.

P.I.X. [T. D. Hallam]. *The Spider Web; the Romance of a Flying-Boat War Flight.* Edinburgh, 1919.

[Plath, C., firm, Hamburg]. *C. Plath, 1862–1962.* Hamburg, 1962. In English.

Plath, C., and F. Schilly. "Sextant à périscope avec compas associé." *Revue hydrographique* 9 (May 1932): 235–37.

Polk, George W. "Instruments Used by the World Flyers." *U.S. Air Services* 10 (Feb. 1925): 31–34.

Pollacchi, P. "La carte aéronautique du Service géographique de l'armée." *La revue aérienne* 4 (25 Apr. 1911): 201–03.

Poor, Charles Lane. "Recent Advances in the Art of Navigation." *Aerial Age Weekly* 9 (26 May 1919): 528–29.

Post, Wiley, and Harold Gatty. *Around the World in Eight Days; the Flight of the Winnie Mae.* New York, 1931.

Prevost, A. "Ascension du 'Touriste.'" *Aéronautique* 3 (Supplement, Jan. 1904): 13–14.

Raleigh, Walter Alexander. *The War in the Air; Being the Story of the Part Played in the Great War by the Royal Air Force.* 6 vols. in 9. 6 vols. text, 2 vols. maps, 1 vol. appendixes. All volumes other than 1 by H. A. Jones. Oxford, 1922–37.

Ranney, H. F. "Maps Again Have Caught Up with Progress." *U.S. Air Services* 11 (Mar. 1926): 41–42.

Rasch, F. "Luftfahrerkarte und Orientierungszeichen." *Deutsche Luftfahrer-Zeitschrift* 16 (21 Feb. 1912): 83–85.

Reece, Robert H. *Night Bombing with the Bedouins.* Boston, 1919.

Renesee, v. "Gedanken über die Ausbildung eines Beobachtungs-Offiziers." *Deutsche Luftfahrer-Zeitschrift* 17 (19 Feb. 1913): 87–88.

Robins, A. W. "Materiel Division Activities for the Year 1937." *Air Corps News Letter* 21 (1 Feb. 1938): 3.

Robinson, Douglas Hill. *The Zeppelin in Combat; a History of the German Naval Airship Division, 1912–1918.* London, 1962.

Robinson, James. "Directional Wireless in Air Navigation." *Automotive Industries* 41 (13 Nov. 1919): 958, 963.

Robinson, Walter D. Review of Émile Guyou, *Nouvelles tables de navigation. Scientific American Supplement* 78 (12 Sep. 1914): 167.

Rogers, Francis M. *Precision Astrolabe; Portuguese Navigators and Transoceanic Aviation.* Lisbon, 1971.

Ross, Raymond L. "Maps for Aviators." *Scientific Monthly* 31 (Aug. 1930): 126–28.

Rotch, Abbott Lawrence, and Andrew Henry Palmer. *Charts of the Atmosphere for Aeronauts and Aviators.* London and New York, 1911.

Rothgiesser, Georg. "Wegweiser für Luftschiffer." *Deutsche Zeitschrift für Luftschiffahrt* 14 (15 June 1910): 17–18.

Rowe, A. P. "Aerial Navigation." *Royal Aeronautical Society Journal* 27 (Sep. 1923): 450–58.

———. *One Story of Radar.* Cambridge, England, 1948.

Russell, Henry Norris. "On the Navigation of Airplanes." *Astronomical Society of the Pacific Publications* 31 (June 1919): 129–49.

"Sailor." "La boussole et les cartes en aéronavigation." *Aérophile* 18 (15 July 1910): 326–27.

Salmet, Henri. "My Paris Flight." *Flight* 4 (16 Mar. 1912): 234–35.

Sanford, R. L. "Direction Instruments; the Testing and Use of Magnetic Compasses for Airplanes." U.S., NACA, Report #128, Part 2. In *7th Annual Report, 1921*, pp. 612–15. GPO, 1923.

Sassoon, Philip Albert G. D. *The Third Route*. Garden City, N.Y., 1929.

Saunière, E. J. "Ascension du ballon 'L'alliance' le 25 janvier 1902." *Aéronautique* 1 (1902): 10–14.

Sayre, Daniel. "First 1/2 Million Miles of Pacific Progress." *Aviation* 35 (Dec. 1936): 15–17, 43–44.

———. "Pacific Preview; Pan American Airways Forms a New Division." *Aviation* 34 (Mar. 1935): 79–82.

Scheller, O. "Flugzeugsteuerung bei unsichtigem Wetter." *Elektrotechnische Zeitschrift* 50 (7 Feb. 1929): 191–92.

Schwarzschild, K. "Tafeln zur astronomischen Ortsbestimmung, von A. Kohlschütter." *Deutsche Luftfahrer Zeitschrift* 18 (7 Jan. 1914): 16.

Scott, C. L. "By Flying Boat to India." *Royal Aeronautical Society Journal* 33 (Sep. 1929): 812–28.

Scott, G. H. "Airship Piloting." *Flight* 12 (9 Dec. 1920): 1254–58.

Scott, P. T. W. "Flight Log." *Aero Digest* 27 (Dec. 1935): 18–20, 72.

Scott, Riley E. "Dropping Bombs from Flying Machines; the Aeroplane as an Offensive Weapon of Warfare." *Scientific American* 105 (28 Oct. 1911): 388–89.

Searle, Frank. "The Requirements and Difficulties of Air Transport." *Flight* 13 (24 Nov. 1921): 789–90. Abstract.

Shangraw, Clayton C. "Radiobeacons for Trans-Pacific Flights." *Institute of Radio Engineers Proceedings* 16 (Sep. 1928): 1203–35.

Shute, Nevil [Nevil Shute Norway]. *Slide Rule; the Autobiography of an Engineer*. New York, 1954.

Siegsfeld, H. Bartsch v. "Astronomische Positionsbestimmungen im Frei-Ballon." *Zeitschrift für Luftschiffahrt und Physik der Atmosphäre* 17 (1898): 2–11.

Skerrett, Robert G. "Radio Compass and Navigation." *Scientific American* 125 (6 Aug. 1921): 89, 105.

Smith, Ross. *14,000 Miles through the Air*. New York, 1922.

Smith-Rose, R. L. "The Equi-Signal Zone Radio Beacon and Air Navigation," *Nature* 126 (19 July 1930): 98–100.

———. "Radio Direction Finding by Transmission and Reception." *Nature* 125 (5, 12 Apr. 1930): 530–32, 568–69.

Sönnichsen, Theo Emil. *Navigation und Seemannschaft im Seeflugzeug; ein Handbuch für Marineflieger*. Berlin, 1918.

Sperry, Elmer Ambrose. "Aerial Navigation over Water." Chapter 16 in Henry Woodhouse, *Textbook of Naval Aeronautics*, New York, 1917.

Squier, George O. "Present Status of Military Aeronautics." *Flight* 1 (22 May 1909): 304.

Starling, S. G. "The Equilibrium of the Magnetic Compass in

Aeroplanes." *London, Edinburgh, and Dublin Philosophical Magazine and Journal of Science,* 6th series, 32 (Nov. 1916): 461–76.

Steffen, Hans, and Otto Bertram. *Orientierung auf Überlandflügen auf Grund praktischer Erfahrungen.* Berlin, 1912.

Stewart, C. J. "Determination of Position in High Latitudes, with Particular Reference to Aircraft Observation." *Nature* 128 (8 Aug. 1931): 212–14.

———. "Instruments." *Air Annual of the British Empire, 1931–32,* pp. 209–19.

Stewart, John Q. "Correction Table for the Effect of Coriolis Acceleration on Bubble Sextant Altitudes." *Review of Scientific Instruments* 11 (Oct. 1940): 310–11.

Stewart, Oliver. *Flying as a Career; a Popular Guide for All Proposing to Obtain a Position in the New Industry as a Pilot, Navigator, or Aircraftman.* 2d ed. London, 1934.

Stuart, A. H. "An Alignment Chart for Obtaining Heights from Observations of Pressure and Temperature." *Aeronautical Journal* 20 (Oct. 1916): 157–61.

Suplee, Henry Harrison. "Methods of Determining Aeroplane Altitudes." *Cassier's Magazine* 39 (Feb. 1911): 329–36.

Surgeoner, David Hay. *Aircraft Radio.* London, 1938.

Süring, Reinhard, ed. *Moedebecks Taschenbuch zum praktischen Gebrauch für Flugtechniker und Luftschiffer.* 3d ed. Berlin, 1911.

——— and K. Wegener, eds. *Moedebecks Taschenbuch für Flugtechniker und Luftschiffer.* 4th ed. Berlin, 1923.

Sykes, Frederick Hugh. "Commercial Aviation in the Light of War Experience." *Flight* 11 (16 Jan. 1919): 84–88.

———. *From Many Angles; an Autobiography.* London, 1942.

———. "Some Empire Aspects of Aviation," address to Australian and New Zealand Luncheon Club. *Flight* 11 (27 Feb. 1919): 271.

Tate, George, Jr. "Improving Airspeed Accuracy by Pitot Tube Location." *Aviation* 32 (Nov. 1933): 339.

Taylor, Patrick Gordon. *The Sky Beyond.* Boston, 1963.

Tear, J. D. "Electromagnetic Compass Systems for Aircraft." In International Civil Aeronautics Conference, Washington, 1928, *Papers Submitted,* pp. 471–73. GPO, 1928.

Tevis, M. "Return Flight of Carrier Pigeons; Novel Theory Advanced—Hints for Aviators." *Scientific American Supplement* 72 (16 Dec. 1911): 391–92.

Thomas, Lowell Jackson. *The First World Flight; Being the Personal Narratives of Lowell Smith, Erik Nelson, Leigh Wade, Leslie Arnold, Henry Ogden, John Harding.* Boston, 1925.

Thompson, F. L. "The Measurement of Air Speed of Airplanes." U.S., NACA, Technical note #616, Washington, 1937. Mimeographed.

Thurlow, Thomas L. *Celestial Avigation.* Rockwell Field, Calif.,

1934. Mimeographed. 44 pp.
———. "Navigation of the Howard Hughes Around the World Flight." *Air Corps News Letter* 21 (1 Nov. 1938): 1–2.
Thurston, A. P. "The Measurement of Air Speed." *Flight* 6 (8, 15 May 1914): 498–99, 521–22.
Toft, Louis. "Effect of the Wind on Flight Speeds." *Flight* 2 (12 Feb. 1910): 108–09.
Towers, John H. "The Great Hop." *Everybody's Magazine* 41 (Nov. 1919): 9–15, 74, 76, 78.
Trenchard, Hugh M. "The Work of the I. A. F." *Flight* 11 (9 Jan. 1919): 52–53.
Treusch von Buttlar-Brandenfels, Horst Julius L. O. *Zeppelins over England*. Translated by Huntley Paterson. New York, 1932.
Tymms, Frederick. "Practical Navigation of Aircraft." *Royal Aeronautical Society Journal* 29 (May 1925): 202–27, 233–39.
U.S. Army. Division of Military Aeronautics. *Technical Orders* #1–17. Dayton, Oct. 1918–June 1920.
U.S. Commerce Department. Aeronautics Branch. *General Airway Information*. Airway bulletin #1, 1 Sep. 1932. GPO, 1932.
———. *Report of Airway Marking Committee*. 23 Jan. 1929. GPO, 1929. 10 pp.
U.S. Federal Board of Surveys and Maps. "Report of the Committee on Aerial Navigation Maps. Jan. 8, 1929; amended Apr. 9, 1929; adopted Nov. 12, 1929." Mimeographed, 3 pp.
U.S. National Advisory Committee for Aeronautics. *1st Annual Report, March 3, 1915–June 30, 1915*. Published as 64th Congress, 1st Session, Senate Documents, vol. 7.
———. *3d Annual Report, 1917*. GPO, 1918.
U.S. Navy Department. Hydrographic Office. *Aircraft Navigation Manual*. H. O. publication #216. GPO, 1940.
———. Hydrographic Office. *The American Practical Navigator, Being an Epitome of Navigation and Nautical Astronomy*, by Nathaniel Bowditch. Revised by Lt. G. W. Logan, USN. H. O. publication #9. GPO, 1903.
———. Hydrographic Office. *Navigation Tables for Mariners and Aviators*. H. O. publication #208. By Lt. Cmdr. J. Y. Dreisonstok. GPO, 1928.
———. Operations: Aviation. *Pocket Manual for Seaplane Pilots*. 1918. 27 pp.
———. Hydrographic Office. *Tables of Computed Altitude and Azimuth*. H. O. publication #214, 9 vols. GPO, 1936–46.
U.S. War Department. Air Corps Tactical School. *Aerial Navigation*. Maxwell Field, Ala., Nov. 1931.
———. Division of Military Aeronautics. *Aerial Navigation. Part I. The Compass. Part II. The Map*. GPO, 1918.
———. Air Service. *Aircraft Instruments, Operation and Maintenance*. Air Service Training Regulation 1440-50. McCook Field, Ohio, 23 Jan. 1925.
———. Office of the Director of Air Service. *Air-Speed Meters*

for Aerial Navigation. War Department document #918. GPO, 1919.

———. Air Corps Tactical School. *Bombardment Aviation Course.* Langley Field, Va., Feb. 1931.

———. Air Service Tactical School. *Bombardment Course, 1924–1925.* Langley Field, Va., n.d.

———. Air Corps Tactical School. *Bombardment Textbook.* Maxwell Field, Ala., 1 Nov. 1935.

———. Air Corps. *Celestial Air Navigation,* TM 1–206. 4 Mar. 1941.

Uttmark, Fritz E. "A New System of Navigation and Nautical Astronomy." *Scientific American Supplement* 81 (17 June 1916): 396–98.

van Vloten, P. van Braam. "L'éclairage des routes aériennes." In Cinquième congrès international de la navigation aérienne, The Hague, 1930, [*Rapports*], 1: 4–20. The Hague, 1931.

Voigt, Oskar. *Ein Beitrag zur astronomischen Ortsbestimmung im Luftfahrzeuge.* Vortrag gehalten am 14. Januar 1911 im Kaiserlichen Aero-Club. Berlin, 1911. 15 pp.

———. "Zur astronomische Ortsbestimmung." *Wiener Luftschiffer Zeitung* 10 (1 Jan. 1911): 8–11.

Volens, Raoul [pseud.]. "Phares pour la navigation aérienne." *Aérophile* 22 (1 June 1914): 250–52.

Volmerange. "La navigation aérienne en France; ses méthodes actuelles et ses applications en particulier à la photographie." In International Congress of Aerial Navigation, 2d, London, 1923, *Report,* pp. 748–56. London, 1923.

von Baumhauer, A. G. "Sphero-Triangulator." In Cinquième congrès international de la navigation aérienne, The Hague, 1930, [*Rapports*], 1: 737–41. The Hague, 1931.

Vorreiter, Ansbert, and Hans Boykow, eds. *Volamekum; Handbuch für Luftfahrer (Ballon, Flugzeug, Luftschiff) nach den neuesten Erfahrungen und Bestimmungen.* Munich, 1912.

Voyer, J. "Note sur la vitesse des navires aériens par rapport au sol." In Congrès international d'aéronautique, IVe, Nancy, 1909, *Procès-verbaux; rapports & mémoires,* pp. 59–66. Paris, 1909.

Wallace, Graham. *The Flight of Alcock & Brown, 14–15 June, 1919.* London, 1955.

———. *R.A.F. Biggin Hill.* London, 1957.

Walter, Louis Heathcote. *Directive Wireless Telegraphy; Direction and Position Finding; the Theory and Practice of Directive Wireless Transmission and Reception . . . .* London, 1921.

Warlick, William Walter. *Naval Aviation; a Textbook for the Instruction of Midshipmen in the Department of Seamanship, United States Naval Academy.* Annapolis: United States Naval Institute, 1925.

Warner, John A. C. "Direction Instruments; Aircraft Compasses—Description and Classification." U.S., NACA, Report #128, Part 3. In *7th Annual Report, 1921,* pp. 616–45. GPO, 1923.

Waters, David W. *The Art of Navigation in England in Elizabethan and Early Stuart Times.* New Haven, 1958.

Watson, Fletcher G., and Henrietta H. Swope. "Loran; the Planning, Growth, and Use of Loran, the New Long-Range Aid to Navigation." *Navigation; the Journal of the Institute of Navigation* 1 (Mar. 1946): 11-15.

Watson, Mark Skinner. *Chief of Staff; Prewar Plans and Preparations.* In the subseries, "The War Department," in the series, "United States Army in World War II." Washington: Department of the Army, 1950.

Webster, Charles, and Noble Frankland. *The Strategic Air Offensive against Germany.* Vol. 1, *Preparation*, and vol. 4, *Annexes and Appendices*. In "History of the Second World War, United Kingdom Military Series," edited by James Butler. London, 1961.

Wedemeyer, A. "Astronomische Ortsbestimmung." In *Moedebecks Taschenbuch für Flugtechniker und Luftschiffer*, edited by Reinhard Süring and K. Wegener, pp. 330-49. 4th ed. Berlin, 1923.

Weems, Philip Van Horn. "Accuracy of Bubble Sextant Observations." *Aero Digest* 28 (Jan. 1936): 65-66.

———. "Air Navigation; Finding Your Way in the Air." *Aviation* 36 (June 1937): 26-27, 76; (July 1937): 22-23, 68, 71; (Sep. 1937): 38-39, 76; (Dec. 1937): 28-29, 74, 77; 37 (Jan. 1938): 28-29.

———. "Analyse des erreurs dues aux accélérations de la bulle." *Revue hydrographique* 13 (May 1936): 78-83.

———. "Dalton Air Navigation Computer Type E-1A." *Aero Digest* 28 (June 1936): 52-53.

———. "Flight of the 'Tingmissartoq'; 1933 Survey Flights of the Lindberghs." *Aviation* 33 (Feb., Apr. 1934): 33-36, 102-05.

———. "Gatty Ground Speed Meter." *Aero Digest* 26 (Jan. 1935): 47.

———. *Line of Position Book; a Short Accurate Method Using Ogura's Altitude Tables and Rust's Modified Azimuth Diagram.* Annapolis, 1928.

———. "Mark VII Navigational Computer for Aircraft." *Aero Digest* 26 (May 1935): 28.

———. *The Secant Time Sight.* Annapolis, 1944.

———. *Star-Altitude Curves for Obtaining a Fix, Given Only the Greenwich Sidereal Time and the Observed Altitudes of Two Stars. . . . Latitude 30° to 41° North.* San Diego, 1928.

——— and Thomas L. Thurlow. "Earth's Rotational Effect on the Bubble Sextant." *U.S. Naval Institute Proceedings* 66 (Oct. 1940): 1420-23.

Wegener, Alfred. "Astronomische Ortsbestimmung im Luftballon." *Illustrierte aeronautische Mitteilungen* 10 (Apr. 1906): 116-21.

Weir of Eastwood, Lord. "Some Developments in Aircraft Design and Application during the War." *Flight* 11 (7 Aug. 1919): 1048-52.

Wellman, Walter. *The Aerial Age;*

a Thousand Miles by Airship over the Atlantic Ocean; Airship Voyages over the Polar Sea; the Past, the Present, and the Future of Aerial Navigation. New York, 1911.

Wentworth, R. Preston. "The Lighthouse for Aerial Navigation." U.S. Air Service 4 (Oct. 1920): 17–18.

Wenz, Émile. "Mesure des hauteurs atteintes par les aéroplanes." Aérophile 18 (1 Jan. 1910): 11–12.

Westervelt, G. C., H. C. Richardson, and A. C. Read. The Triumph of the NC's. Garden City, N.Y., 1920.

Westover, Oscar. "1936 Passes in Review; the Army." Aviation 36 (Jan. 1937): 64.

White, James Dundas. Steering by the Stars for Night-Flying, Night-Marching, and Night Boat-Work, between Latitudes 40° N. & 60° N. . . . London, 1916.

White-Smith, H. "The Operation of Civil Aircraft in Relation to the Constructor." In Great Britain, "Proceedings of the Air Conference, 1920." Cmd. 1157, Parliamentary Papers, 1921, 8: 34–50.

Wilke. "Speed Measuring Instruments for Aeroplanes." Scientific American Supplement 87 (28 June 1919): 414. Translated from Der Motorwagen.

Wilkins, Hubert. "Some Aspects of the Graf Zeppelin's Flight around the World." Aviation 27 (21 Sep. 1929): 594–600.

Williamson, G. W. "Instrument Progress and Development." Air Annual of the British Empire, 1938, pp. 177–95.

Wimperis, Harry Egerton. "Air Navigation." Encyclopaedia Britannica, 12th ed., 30: 41–45.

———. "Air Navigation; the Most Important of the Unsolved Problems Relating to Aviation." Aeronautics 16 (8 May 1919): 482–87. Also Aeronautical Journal 23 (Aug. 1919): 445–61.

———. "Navigational Equipment for Long-Distance Flights." In International Congress of Air Navigation, 2d, London, 1923, Report, pp. 756–64. London, 1923.

———. A Primer of Air Navigation. London, 1920.

———. "Some Recent Developments in Aircraft Instruments." Royal Aeronautical Society Journal 28 (Jan., Feb. 1924): 3–34, 45–48.

"Wing Commander." Bombers' Battle; Bomber Command's Three Years of War. London, 1943.

Woodhouse, Henry. Textbook of Naval Aeronautics. New York, 1917.

Young, D. C. "Aerial Highways and Their Marking." American City 39 (Nov. 1928): 127–29.

Zahm, Albert Francis. "Development of Air Speed Nozzles." U.S., NACA, Report #31. In 4th Annual Report, 1918, pp. 322–34. Published as 65th Congress, 3d session, Senate Documents, vol. 2. GPO, 1920.

INDEX

Aberdeen Proving Ground, 175
Abney spirit level, 109
accidents related to navigation, 43–44, 65, 91–92, 102, 110, 113–15, 126, 132–33, 136–37, 167, 175, 180, 233 n. 11
Adcock, Frank, 121–22
Admiralty, British, 43, 45, 70, 101, 160
Advanced Avigation Training Units (U.S. Army Air Corps), 181
Advisory Committee for Aeronautics (British), 46–47, 70
Aéra (firm, Paris), 134
aero club: earliest, for ballooning, 17, 19; of France, 21; of America, 21; of Pau, France, 56
Aeronautical Exposition, Paris (1910), 43
Aeronautical Society of Great Britain, 52, 102, 115, 176
Ageton, Arthur A., 161–62
Air Commerce Act of 1926 (U.S.), 117, 124
Air Conference of the British Empire, 1st, London (1920), 142
Air Corps Board (U.S.), 172
Air Corps Tactical School (U.S.), 172–75, 178–83, 185
aircraft carrier, 169
Air France, 131, 167
Air Line Pilots Association, 126
airlines and airliners, 102–03, 113–15, 120–21, 125–29, 131–33, 139, 141, 143, 146–48, 167, 199 (photos, 61, 104). *See also* names of individual airlines.
airmail, 102, 113, 116, 118, 120, 123–24, 131, 181
Air Ministry (Great Britain), 111, 160

airplanes (by type and name), 209; *Albatros*, 49; ANT-25, 139, 148; AT-7, 189 (photos, 44, 157, 192); B-9, 173, 181; B-10, 173, 181, 193; B-15, 194; B-17, 142, 168, 173, 181, 194–96, 201 (photo, 195); B-18, 184; Boeing 247, 126; Boeing 314, 149, 156, 168 (photo, 61); Breguet *Question Mark* (photo, 163); BT-2B (photo, 144); Caproni 49, 95, 98; Curtiss *America* 100 (photo, 59); DH-4 (photo, 101); F.2a, 100; Fokker Trimotor, 136 (photos, 61, 75, 112, 154); "Giants," 78, 93, 95, 98 (photo, 94); Gotha, 95, 98; Handley Page 0/400, 65, 95, 98, 110 (photo, 104); Handley Page V/1500, 95, 102, 105 (photo, 95); Martin 130, 156; NC, 104–05, 109–10, 123, 127, 147 n, 230 n. 55; O2U (photo, 170); PBY-1, 156; Sikorsky 42, 156; *Southern Cross*, 148; *Spirit of St. Louis*, 137 (photo, 136); Vickers Vimy, 104; XPBY2Y, 156
airships (by name and serial number), 209; *America*, 60; *Graf Zeppelin*, 131, 137, 145; *Hindenburg*, 131; *L 3*, 89; *L 4*, 89; *L 20*, 91; *L 48*, 91; *L 70*, 93; *LZ 126*, 143; *Norge*, 139; *R 34*, 104–05, 143
airsickness, 113
airspeed indicator: needed for safe operation, 47–48; before World War I, 47–50, 66; during the war, 73–74; postwar, 134; converting indications of, to true

INDEX

airspeed, 67, 134, 182
Alcock, John, 104, 109, 165
almanac: air, 158–60, 199; nautical, 16, 32, 139, 156–57, 160 (photo, 158)
altimeter, absolute: acoustical, 27, 142, 143 (photo, 144); capacitance, 142; electrical conductors suspended from balloon, 27; gravity, 27; need for, 142; optical, 143; radio, 143–45
altimeter, pressure: mercury barometer as, 26; pre-World War I, 26–27, 50, 53; recording, 27, 143; development in World War I, 74–76, 78; postwar, 142
ambiguity, 180°, 84–85, 122–23
American Society of Mechanical Engineers, 126
amplifier tube, 83, 87–88
Amundsen, Roald, 139
Andrée, Salomon August, 30
Andrews, Frank Maxwell, 184
anemometer as airspeed indicator, 48–50, 66, 74
aneroid. *See* altimeter, pressure
antenna, 83–85, 87, 121–23, 125, 127 (photos, 61, 82)
antiaircraft fire, 65, 74–75, 91, 97–99, 174
aptitude test, 193
Aquino, Francisco Radler de, 106, 161
Archbold Expedition, 156
Aries, first point of, 16, 34 n
Arnold, Henry Harley, 175, 178, 180, 183, 186–87, 193, 197, 200 (photo, 81)
artificial horizon, 73 (photo, 144)
artificial landmarks: in marine navigation, 4; for early air races, 55–56; schemes for permanent, 56–57, 117–18; for overwater patrol, World War I, 100; provided by line of destroyers across North Atlantic for NC flying boats, 110. *See also* beacon (light)
artillery spotting, 63–64, 82, 109, 177
assumed position, 12–13, 33, 161
astro compass. *See* sun compass
astrodome, 151, 155–56, 189, 199 (photos, 61, 157)
astrograph, 165–66
astronomical triangle. *See* celestial triangle
Ault, James Percy, 106–07
autopilot, 149, 174, 187
averaging celestial observations, 31, 106, 149–53
avigation, 240 n. 9

Badin airspeed indicator, 73–74
Baker, Thomas Yeomans: air sextant, 108–09, 226 n. 12; navigating machine, 109, 165
Balchen, Bernt, 233 n. 11
balloons: early history of, 17; controlling altitude of, 24
Bamberg, Carl (firm, Berlin): Double Compass, 37; selenium compass, 71
Barksdale Field, Shreveport, 189–90
Bausch & Lomb (firm, U.S.), 133
beacon (light): pre-World War I, 57–58; World War I, 64, 97–98; postwar, 118–20, 124, 177 (photo, 119)
bearing compass, 4, 7, 23–24, 26, 37, 42, 101
Bellini-Tosi directional receiver, 84–85
Bell Laboratories, Bell Telephone, 127, 133
Bellonte, Maurice, 149, 163–64
Bennett, Donald Clifford Tyndall, 145–46, 160, 176
Bennett, Floyd, 139
Bennett, G. T., 70–71
Biggin Hill, RAF Station, 88, 105
binnacle, 7
blackout, 65, 91, 96, 99, 197
Blériot, Louis, 42

Blue Hill Observatory, Hyde Park, Mass., 38
board of officers, 177–78
Bockholt, Ludwig, 93
bombardier (as distinct aircrew member): none in World War I, 77; U.S. Army Air Corps, 184–85, 193, 196
bombing accuracy and navigation, 88–90, 94, 99–100, 171, 174–75, 194, 197, 224 n. 67
bombing doctrine, long-range, 94–95, 169–75, 180, 201
bombing range, 196–97
bombsight: need for, 76; in World War I, 76–78, 96, 99, 221 n. 25; postwar, 171–72, 174, 197
Booth, Lionel Barton, 107
Boy Scouts of America, 117
Bradley, Follett (photo, 81)
Brand, C. J. O., 104
Brill, Alfred, 35, 165
British Airways, 167
British Association, 21
Brown, Arthur Whitten, 104, 109–10, 165
Bumstead, Albert H., 138–39
Bureau of Aircraft Production (U.S.), 105
Bureau of Standards (U.S.), 123–24, 135, 141, 148, 153
Butenschön sextant, 30–31
Bygrave, L. C., slide rule, 111, 160 (photo, 112)
Byrd, Richard Evelyn, 107, 109, 136–39, 145, 199, 233 n. 11

calibration of instruments, 183
Calshot, RAF Station, 176
Cambridgeshire, 53
camera obscura, 96
Campbell, G. R. C., 70–71
Carnegie Institution, 106–07
celestial navigation: in marine navigation, 11–15; in balloons, 29–36; in airplanes before World War I, 58, 62, 219 n. 49; in airships, 60–62, 92–93; in airplanes after World War I, 105–13, 128, 133, 147–66, 178, 181, 183–87. *See also* almanac; astrodome; averaging celestial observations; sextant, artificial horizon; sextant, marine
celestial triangle, 11–13, 16, 31–32, 36, 160–61, 164
Chamberlin, Clarence Duncan, 136
chart, aeronautical: need for, 18–19; topographical charts as, 18, 20, 22, 53; early, 20–22, 52–53; photographs as, 21–22, 116; use in computing groundspeed, 27; mounted on rollers, 52–53; basis for numbering artificial landmarks, 56; in World War I, 64–65; postwar, 116–17, 239 n. 70
chart, marine: in pilotage, 4; water depth and bottom on, 5; variation on, 7, 37; tidal streams and currents on, 9–10
chart board, 67
checkpoint, 54
Chichester, Francis Charles, 160, 163, 176
chronometer, 14–15
Clipper. *See* Pan American Airways; airplanes (by type and name): Boeing 314, Martin 130, Sikorsky 42
cloud horizon, for measuring height of celestial body, 106–08
Coast and Geodetic Survey (U.S.), 22, 117
coastline effect, 122
Cody, S. F., 43, 55, 216 n. 7
College Park, Maryland, 125
Collins, E. B., 161
combat crew, 184–86
Commerce Department (U.S.), 117, 124–26
compass (directional instrument). *See* bearing compass; directional gyro; gyro compass; magnetic compass; sun compass

Index

274

compass points, 5–6, 44, 85
compensation: defined, 7; in balloons, 23, 37; in airplanes, 43–46, 54, 65, 97, 101, 109
cone of silence, 125
Congress of the International Aeronautical Federation, 3d, Brussels (1907), 20
Conneau, Jean, 46, 54
Constance, Lake, 42, 95, 131
Copernicus, 14
Coriolis's acceleration, 153–54
courbe du chien, 40
course indicator: on early airplane compasses, 44–45; on remote compasses, 71, 135
Coutinho, Carlos Viegas Gago, 111–13, 155
crew coordination, 16, 31, 58–60, 77–78, 106, 111–13, 132, 149, 184
Curtiss, Glen, 62, 100
curve of pursuit, 41

Daloz compass, 51
dead reckoning: defined, 3–4; to identify landmarks, 4, 55; on ships, 9–11, 39; in balloons, 18, 22–23; in airships, 51, 60–62, 90; in airplanes, 87, 97–98, 101, 109–11, 126, 142, 147, 166, 179, 181, 183, 197
dead-reckoning computers, 40, 66–67, 97, 139–40, 182, 220 n. 12, 240 n. 70 (photos 68–69, 72, 128, 182)
declination: defined, 14, 34; tabulated in almanacs, 16, 158
Deutsche Lufthansa, 131, 139, 167
Deutsche Luftschifferverband, 20
Deutsche Versuchsanstalt für Luftfahrt (DVL), 155
deviation: defined, 7; memory crutches for using, 8; in balloons, 23, 37; in airplanes, 43–45, 71, 97; in airships, 60; in directional radio, 85; measured in-flight with periscopic sextant, 155. *See also* compensation
dip: with marine sextant, 15, 62, 108; magnetic, 37, 70
directional gyro, 137–38, 149 (photo, 144)
directional radio, airborne: antenna wrapped around wings, 87, 120, 123, 127; rotatable coil, 87–88, 105, 110, 120, 123, 127, 230 n. 55; post-World War I, 113, 120, 122–23, 127; radio compass, 122, 127, 194 (photo, 190)
directional radio, ground: early proposals for use in navigation, 81–82; first used in marine navigation, 84–85; in World War I, 86–87, 90–94, 101–102; errors, 88, 91, 101–102, 121–22; postwar, 120–26, 132, 231 n. 59. *See also* net, ground radio; radio range
Division of Military Aeronautics (U.S. Army), 73
double-drift wind, 78–80, 141
Douhet, Giulio, 171
Dreisonstok, J. Y., 161, 165
drift, killing the, 41
drift marks painted on aircraft, 111 (photo, 112)
driftmeter: in balloons, 25–28; in World War I, 77–80, 101; postwar, 109, 128; difficulty of pilot using, 58; Bureau of Standards, 141; Crocco, 79–80; Daloz compass, 51; Gatty, 141–42 (photo, 140); gyro-stabilized, 142, 194; le Prieur Navigraph, 140–41; Pioneer, 80; Sperry, 80, 101; Wimperis, 79–80, 101, 105; Zeiss Quo Vadis, 139–40. *See also* bombsight

Earhart, Amelia, 132, 138, 156
Eckener, Hugo, 143, 145
Eiffel Tower, 56
Eilson, Carl Ben, 137

INDEX

275

Ellsworth, Lincoln, 139
emergency landing, 18, 64, 92
Emmons, Delos Carleton, 188
Engineering Division, U.S. Army Air Service/Corps, 135, 150, 176–77. *See also* McCook Field; Materiel Division, U.S. Army Air Corps
English Channel, 28, 42, 54–56, 80, 86, 98–100
equisignal, 123–24, 133
evasive action, 96, 98

Fairchild Aerial Camera Corp. (U.S.), 127, 150, 152, 164–65
Farmingdale, Long Island, 197
Favé, L., 33–35, 165
feed-back circuit, 127
fix: celestial, 106, 109, 147, 158, 183; hybrid, 109, 167; map reading, 55, 89, 98; radio, 86, 90; running, 5, 101
flares: floating, for measuring drift over water, 24, 90, 142, 167; parachute, for night map reading, 89, 174–75, 197
Fleuriais gyroscopic sextant, 29
flight plan, 97
Fokker, Anthony Herman Gerard (photo, 54)
formation flying, 96, 99, 172, 174, 193–94
Frankfurt a. M. International Aeronautical Exposition (1909), 20, 35
frequency meter, 144

Gardner, Grandison, 179–80, 183
Gatty, Harold, 137–38, 141–42, 162–63, 199
General Aeronautical Co., Ltd. (Great Britain), 74
General Electric Co. (U.S.), 70, 117, 133
General Headquarters Air Force (U.S. Army), 181, 184, 186–87
Gingrich, John Edward, 161

globe, for reducing celestial observations, 32
Goerz, C. P. (firm, Germany), 133, 139
Grahame-White, Claude, 55
Gray, Paul, 153
great circle route, 145
Greenwich hour angle (GHA), 158–60
Grieve, Kenneth Mackenzie, 108–09
Griffiths, John F., 165–66
ground markings. *See* artificial landmarks
group navigation officer (U.S. Army Air Corps), 186, 200
guide rope, 24, 26, 62
gyro, 29, 73, 78, 128, 137–38, 142
gyro compass, 137–38

Habana Convention on Commercial Aviation, 115
Hagner position finder, 165 (photo, 164)
Hamel, Gustav, 58–59
Harbold, Norris Brown, 188
Hawker, Harry, 108
heading, 7, 39, 41–42, 50, 73, 97–98, 145
homing, 87, 110, 120, 169
homing sense, 42
Hondo Army Air Field, Texas, 191
Hughes, Henry, & Sons (London), 133, 150, 152–54, 160, 216 n. 9
Hughes, Howard, 150 n, 156
Hydrographic Office (USN), 117, 147 n, 158–59, 161, 166: H.O. publication 208, 161; H.O. pub. 214, 161

ignition interference, 83, 110
Imperial Airways, 131, 150
Independent Air Force (RAF), 100, 169
inertial navigation, 128, 199
Institute of Navigation, 151
instrument design, general considerations, 42, 65–66, 73, 83, 134,

145, 148, 165, 198
instruments, location and mounting of: in balloons, 23, 25–27, 37; in airships, 49–50, 60–62, 139; in airplanes, 43–45, 48–51, 53, 77, 83, 139 (photos, 101, 144)
intelligence from radio interception, 86, 88
intercept in celestial navigation, 13, 32–33, 106
intercom system, 59, 77–78, 176
International Air Navigation Convention (1919), 114–17
International Commission for Aeronautical Charts, 20–21
International Commission for Air Navigation, 114–16
International Congress of Aerial Navigation, 1st, Paris (1921), 115
International Congress of Aeronautics: 2d, Paris (1900), 19, 33; 4th, Nancy (1909), 21
International Map of the World, 21
ionosphere, 121
"iron compass," 118

Kelly Field, San Antonio, 189, 192
Kingsford-Smith, Charles Edward, 137
knot, derivation of term, 9
Kollsman Instrument Co. (Brooklyn), 134, 150
Kreusi, Geoffrey, 127
Kuter, Laurence Sherman, 185

land effect, 122, 125–26
Langley Field, Virginia, 105–07, 172, 181, 186–87, 193–94, 199
latitude by Polaris, 14, 16, 35
leeway, 9–10
LeMay, Curtis Emerson, 181, 187, 199
Lepetit, A. (firm, Paris), 150
le Prieur, Yves, 140–41
Lindbergh, Charles Augustus, 136–38, 148–49, 159, 163, 233 n. 13

line of position: from visual bearing, 4; from depth sounding, 5; celestial (Sumner), 12–13, 32–34, 106, 109, 111, 113, 149, 164–67, 183, 194; magnetic, 37; radio, 84, 86, 167
Link, Edwin A., and Link Aviation Devices (New York), 134, 148, 151–52, 154, 190–91
log: air, 47; chip, 8–9, 25–26; patent, 9–10, 24, 47, 62; pressure, 9
log book, 9, 62, 114–15, 166
loran, 198
lubber line, 7, 42, 44, 65, 80
Lucas, Keith, 70–71
Lufthansa, 131, 139, 167
lunar supplement to *American Nautical Almanac* (photo, 158)

McCook Field, Dayton, 124, 135, 176–77. *See also* Engineering Division, U.S. Army Air Service/Corps; Wright Field
"magnetic" clouds, 46
magnetic compass: marine, 5–7; in balloons, 23; in early powered aircraft, 42–47, 53–55; in World War I, 63, 70–71, 91, 101 (photo, 74); postwar, 134–38 (photo, 75); Bamberg remote (selenium), 71, 134–35, 137; Campbell-Bennett 6/18 aperiodic, 70–71, 105 (photo, 75); Creagh-Osborne 5/17, 70; General Electric B, 70; Pioneer earth inductor, 134–37, 173, 233 n. 13 (photo, 135); R.A.E. Mk II, 70
magnetic positioning attempted, 36–37
maneuvers, military: French (1910), 21; British (1911), 21, 53; American (1937–38), 195–97
map. *See* chart, aeronautical; chart, marine
map reading: in balloons, 17–18,

22–23, 28; in powered aircraft before World War I, 50–51, 53–55; during World War I, 63–65, 89–91, 96–98, 100; postwar, 104–05, 113–15, 129, 133, 172–73, 176–77, 179 (photo, 191). *See also* pilotage

March Field, Riverside, 181, 183–84, 186

Marconi, Guglielmo, 80, 85, 127

marker beacon, 125–26

Materiel Division (U.S. Army Air Corps), 178, 187, 200

Mather Field, Sacramento, 189

meridional transit, 14

meteorology: of upper air largely unknown before World War I, 28–29, 33, 37, 50; first methodical upper-air observations, 20, 38; anemometers, 48–49; during World War I, 64, 66, 89, 91–92, 97, 102; variation of wind with altitude, and bombing problem, 77; postwar, 115, 143, 145, 174–75, 177, 186, 194, 196, 200–01; in polar regions, 139; use of Coriolis's acceleration, 153

Mitchell, William, 171–72

Moedebeck, Hermann Wilhelm L., 18–20, 22; *Moedebecks Taschenbuch* (1911), 36

Morse code, 57, 97, 123–24, 155, 177

mountain effect, 122, 125–26

National Advisory Committee for Aeronautics (U.S.), 47 n, 50, 161

National Geographic Society, 138

Naval Institute, U.S., 153, 161

navigation: relation of aerial to marine, 3; need for, in commercial aviation, 113–14, 132; international regulation of, 114–16, 121; of transocean airliners by all available methods, 166–67; priority of development in U.S. Army Air Corps, 176, 188, 200; need for, U.S. Army Air Corps (1929), 179–80; in wartime compared to peacetime, 198; pattern of development of aerial, 198–99

navigator as distinct aircrew member, 55, 58–60, 99, 102, 114–15, 120, 147, 172, 183–85, 189, 193–94

navigator training: in World War I, 64–65, 93, 99, 102; postwar civil, 114–15, 141 n, 163, 165, 167, 176; postwar military, 165–67, 175–93 (photos, 186, 190–92)

"Navitrainer" (photo, 190)

net, ground radio, 86, 101, 120–21, 125, 127, 133

night effect, 121–22, 125

night fighter navigation, 65

19th Bombardment Group (U.S. Army Air Corps), 186–87, 196–97

Noonan, Frederick, 132

North Atlantic, early flights across, 35, 60–62, 104–05, 107–10, 131–32, 136–38, 143, 145, 147 n, 148–49, 177 (photo, 59)

Notices to Mariners, 4

null, in radio direction finding, 84–85, 87–88, 122–23

Olympia Air Show, 3d, London (1911), 43

omnirange, 127

Opitz, Wilhelm, 155

oscillation: of aircraft, 149; of balloon, 17–18, 23; of compass, 7, 45, 54, 70–71; period of, defined, 45 n

oxygen, 91, 194

Pacific Ocean, early flights across, 132–33, 137, 148 (photo, 61)

Pan American Airways, 131–33, 138, 142, 145, 149, 153, 156,

159–60, 165, 167, 188–89, 193–94, 199 (photo, 61)
pantograph, 140
parallax: in reading compass, 43; in sextant observations, 15
patrol, overwater, World War I, 65, 80, 100–02
Paulhan, Louis, 55
pilotage: in marine navigation, 3–4; in balloons, 17; in airplanes, 40–41, 50. *See also* map reading
pilot training: in Great Britain, World War I, 64–65; in U.S., postwar, 115, 177
Pioneer Instrument Co. (U.S.), 80, 133–37, 148, 152, 154
pitot tube, 48–50, 73
Plath Co. (Germany), 133, 155–56
plotting instruments: in marine navigation, 16; in airplanes, 67, 166
plotting machines for celestial navigation, 35, 109
polar regions, flights in, 30, 60, 137–39, 148
Ponthus et Therrode (firm, Paris), 30–31
Poor, Charles Lane, slide rule, 106
Porte, J. C., 62
Post, Wiley, 137–38, 141, 162–63
Post Office Dept. (U.S.), 116, 120, 123–24
postwar adjustment in aviation industry, 103
precomputation in celestial navigation, 13, 33–35, 106
pressure pattern, 133, 143, 145
prizes. *See* races
prohibited areas, 57
Public Works Administration (U.S.), 117

races: balloon, 17, 28, 35; Britain to Australia, 103–04, 113; Britain to South Africa, 103–04, 107; *Daily Mail* London-to-Manchester, 55; *Daily Mail* 1st crossing of North Atlantic in 72 hours, 103–04; Dole, 133; European Circuit, 42, 56; German Circuit, 22, 42, 53, 58; Saxony, 42, 53, 58
radar, 145, 198
radio beam. *See* radio range
radio communications, 60, 80–83, 86, 105, 109, 111, 120–21, 125, 223 n. 41 (photos, 81–82)
Radio Corporation of America, 127, 133
radio direction finding. *See* directional radio, airborne; directional radio, ground
radio range, 123–28, 133, 173, 178, 196
radius of action, 95, 100, 179
Rand McNally's *Air Trails Maps*, 116
Randolph Field, San Antonio, 144
rate of timepiece, 15
reconnaissance, 63, 82, 109, 169, 177, 186–87
refraction, 15
Republic (U.S. Army ship), 196
resonance, 45 n
Rex (ocean liner), 195–96
Richard, Jules (firm, Paris), 133
right ascension, 16
Robinson, Douglas Hill, 91
Robinson, James, 87–88
roof markings. *See* artificial landmarks
routes prescribed by governments, 115
Royal Aeronautical Society, 52, 102, 115, 176
Royal Aircraft Establishment/Factory, Farnborough, 46, 47 n, 70, 107
RPM indicator for measuring speed, 9
Russell, Henry Norris, 106–07
Ryneveld, Pierre van, 104

sailing directions, 4

St. Hilaire, Marc, 13, 32–33, 106
Scott, Riley E., 76
sensing circuit (DF), 122–23
sextant, artificial horizon: averaging, 150–53; early air bubble, 30–31, 106, 113; gyroscopic, 29, 107, 226 n. 12; mercury, 29–30, need for, 29; pendulum, 29–31, 33, 62, 106; Booth, 107; Bureau of Standards, 148 (photo, 108); Byrd, 107; Coutinho, 111, 113, 148; Fairchild, 150, 152; Favé-Lepetit recording, 150–51; Hughes, 150, 152–53; Link (photo, 152); Pioneer, 152; Plath periscopic, 155–56; R.A.E., 107, 111, 148; Schwarzschild, 31; SOLD (Plath), 107, 148; Willson, 106–07
sextant, marine, 5, 11, 15, 62, 93, 106, 109, 113 (photo, 146)
Simon, Murray, 60–62
Smith, Ross and Keith, 104–05, 113
Smith, S., & Sons (London), 133
South Atlantic, early flights across, 111–13, 131
Sperry, Elmer, and Sperry Gyroscope Co., 78, 80, 120, 127, 134, 137–38, 144, 149
"Spider Web," 100
spin, airplane in, 46, 105
Spitsbergen, 30, 60, 137
Squadron 101 (French night bomber), 98
stall, 48, 50
"Star Curves," 35, 161–63, 165–66
Steffen, Hans, 54, 58–59
Stewart, John Q., 153
subpoint, 11–14
Sumner, Thomas Hubbard, 12–13
sun compass, 138–39
swinging the ship. See compensation
Taylor, Patrick Gordon, 147
Telefunken directional transmitter, 84–85, 87, 93, 120
temperature, free air, 26, 182

textbooks, navigation, 66, 160, 187
Thurlow, Thomas L., 150, 153, 160, 166, 187 (photo, 151)
time, 14–15, 34 n, 35, 157–59
Tinker, Clarence Leonard, 185
Transformators, 36
traverse table, 10, 16, 110
Trenchard, Hugh Montague, 99, 175
Treusch von Buttlar-Brandenfels, Horst Julius J. O., 93
Trinity House, 118
turn indicator: compass used as, 46–47, 104; developed, 73. See also artificial horizon
turning error of compass, 46–47, 70–71, 138
216 Squadron (RAF bomber), 98

Ulm, Charles T. P., 132, 137
University of Miami, 188
Utah (battleship), 195

variation: defined, 7–8; memory crutches for using, 8; attempts to determine position by measuring, 37; displacing lubber line to remove, 65
vector diagram, 39–40, 66–67, 79, 166
venturi tube, 73
vertical, determining the, 27, 78, 137. See also sextant, artificial horizon

Waghorn, David, 152
weather forecasting. See meteorology
weather map flying, 133, 143, 145
Webster, D. L., 106
Weems, Philip Van Horn, and Weems System of Navigation, 141 n, 147 n, 148–51, 153, 156, 158–66, 187, 194 (photo, 146)
Wellman, Walter, 60
Westover, Oscar F., 196
Wilkins, George, 137

Willson, Robert Wheeler, 106–07
Wimperis, Harry Egerton, 79–80, 102, 140, 161
wind forecasting. *See* meteorology
wind-gauge bearing plate, 79
wind star, 78–80, 141
wind triangle, 40–41, 66–67, 79
Wissenschaftliche Gesellschaft für Luftfahrt, 149
working conditions of navigator: in marine navigation, 16; in balloons, 23–24, 32; in airships, 42, 52–53, 91, 139; in airplanes, 42–44, 52–53, 55, 67, 77, 83, 98, 108, 110–11, 120, 154–56, 164, 193, 199 (photo, 170)

Wright Field, Dayton, 127. *See also* Engineering Division, U.S. Army Air Service; McCook Field, Dayton

Yancey, Lewis A., 154–55

Zeiss (firm, Germany), 133, 139–40
Zeppelin: prewar airships, 20, 42; naval patrol, World War I, 100; bombing England, World War I, 65, 85–95, 98, 101, 223 n. 51; postwar, 131, 137, 143, 145. *See also* airships (by name and serial number)

THE AUTHOR

Monte Duane Wright, born 9 September 1930, Comanche County, Texas. Graduated, 1947, Roswell Senior High School, New Mexico. B.A., 1951, M.A., 1953, Baylor University, thesis: "Roosevelt, the State Department, and the Third Reich, 1933-1941." Ph.D., Duke University, 1970, dissertation: "A History of Aerial Navigation to 1941." Publications: "Demography for Historians," *Rocky Mountain Social Science Journal* 7 (Oct. 1970): 1–10; "USAF Tactics against Air and Ground Defenses in SEA, Nov. 68–May 70," Project CHECO report, HQ 7AF DOAC, 25 Sep. 1970, classified; *Science, Technology, and Warfare; the Proceedings of the Third Military History Symposium,* USAF Academy, May 1969 (co-editor; GPO, 1971); *Soldiers and Statesmen; the Proceedings of the Fourth Military History Symposium,* USAF Academy, Oct. 1970 (co-editor; GPO, 1973).

Continuous service in the United States Air Force since 1951. Present temporary rank, lieutenant colonel. Assignments, other than military schools: aircraft observer, radar all-weather interceptor; ground-control intercept director; navigator; mission analysis and plans officer, wing level; instructor in navigation, and currently Associate Professor of History, USAF Academy, Colorado 80840.